Arthur Wisswald Weysse

An Epitome of Human Histology

For the Use of Students in Connection with Lectures and Laboratory Work

Arthur Wisswald Weysse

An Epitome of Human Histology
For the Use of Students in Connection with Lectures and Laboratory Work

ISBN/EAN: 9783337408749

Printed in Europe, USA, Canada, Australia, Japan

Cover: Foto ©berggeist007 / pixelio.de

More available books at **www.hansebooks.com**

AN EPITOME

OF

HUMAN HISTOLOGY

FOR

THE USE OF STUDENTS IN CONNECTION WITH LECTURES AND LABORATORY WORK

BY

ARTHUR W. WEYSSE, A.M., PH.D.

INSTRUCTOR IN BIOLOGY, MASSACHUSETTS INSTITUTE OF TECHNOLOGY
BOSTON, U.S.A.

NEW YORK

LONGMANS, GREEN, AND CO.

LONDON AND BOMBAY

1898

PREFACE.

THE student who has conscientiously followed a course of lectures, laboratory work, or reading, on microscopic anatomy generally finds himself in possession of a great many isolated facts about the minute structure of the body, but with a rather indefinite conception of the relation of these facts to one another and of the subject as a whole. He often finds it necessary to go over his notes or his text-book, eliminating those matters which, though of interest, it is neither necessary nor desirable for him to memorize. This process of selecting and arranging the facts of histology is far from easy, especially for the first-year medical student, who thus often enters on the study of pathology with an inadequate knowledge of the normal structure of the tissues.

It is to overcome this difficulty as far as possible that this epitome has been written. Several points have been kept constantly in mind: first, to present all the facts which are of real importance for the student, — a point often neglected in the "aide-mémoire" and similar books ; second, to express these facts in the briefest and clearest language, omitting all points and phrases which, though they might add interest or literary merit, are not strictly required ; and third, to arrange the facts in such a way that the reader, in considering any organ, may, if he will, actually sketch each part described as he proceeds, and thus make a diagrammatic plan or picture of the entire structure ; this will obviously be possible for those students only who have seriously followed a regular course in histology. It is for such students that this book is intended, and not for those who have neglected their work and are anxious to acquire in a few hours facts enough to pass an examination.

One may see at a glance that this is not a text-book, nor is it intended to replace the text-book. It can, furthermore, only serve its true purpose when used by students who have had laboratory practice as well as lectures in histology, and have thus examined the actual structures.

An acquaintance with histology acquired by lectures and text-books alone is of as little value as an idea of gross anatomy acquired without dissection.

In writing this epitome I have consulted freely the various text-books, a large number of original papers, notes which I took a few years ago while studying abroad, particularly at Berlin and Paris, as well as preparations of the organs themselves, both those in my own collection and many belonging to my friends. It was beyond the scope of this book to discuss mooted points, but I have always endeavoured to give the interpretation which, so far as I know, is accepted by the best authorities. In the case of several structures, however, I have mentioned that there is a diversity of opinion, believing that nothing is more stimulating to the true student than to know that there are still many subjects requiring further elucidation.

A. W. WEYSSE.

Boston, January, 1898.

TABLE OF CONTENTS.

AN

EPITOME OF HUMAN HISTOLOGY.

DEFINITIONS.

Human histology is the science of the tissues of the human body.
A **cell** is a nucleated mass of protoplasm.
A **tissue** is an aggregate of similarly differentiated cells and their products having a common function.
An **organ** is any part of the body having a definite function, and consists of a complex of tissues.

THE CELL.

I. A typical **cell** consists of a **cell membrane**, its protoplasmic contents, the **cytoplasm**, and a **nucleus** consisting of a **nuclear membrane** and its contents, the **nucleoplasm**.

II. **Protoplasm** consists : —

1. **Chemically,** of **C**, about 72 parts ; **O**, 32 ; **H**, 106 ; **N**, 18 ; **S**, 1 ; and a little **P** ; and in the nucleus **Fe** and **Cl** in addition. Some sixteen elements have been discovered in the human body.

2. **Physically,** of the **plasma**, **spongioplasm**, or active portion, which contains granules, the **microsomes**, and is arranged in threads or films holding in its meshes the **chylema**, **hyaloplasm**, or passive portion, which is hyaline.

III. The **nucleus** consists of **chromatin** and **achromatin**, the former including the **nuclear network** and the **nucleolus**, the latter the **linin fibres**, the **nuclear fluid**, and the **nuclear membrane**. In the nucleus also lies the **centrosome**, a deeply stainable body, surrounded by a clear area. During cell-division this appears in the cytoplasm near the nucleus.

IV. **Cell multiplication** consists in a division of the nucleus succeeded by a division of the cytoplasm. **Direct or amitotic** nuclear division is rare in man, — it may occur in the leucocytes. Typical cell division is called **indirect, mitotic,** or **karyokinetic,** and consists of three phases occupying about half an hour in man.

1. The **prophase** is characterized by the following more or less simultaneous changes : —

a. The centrosome passes into the cytoplasm and becomes surrounded by a clear zone and radiations constituting the **attraction sphere;** it then divides, and the two parts move around the nucleus through arcs of 90° in opposite directions.

b. The nuclear membrane and the nucleolus disappear.

c. The nuclear network thickens and forms a **close skein,** consisting often of V-shaped pieces, the **chromosomes,** arranged about a clear area, the **polar field.** A further thickening produces the **loose skein.**

2. The **metaphase** is characterized by : —

a. The **nuclear-spindle,** which consists of threads extending between the two centrosomes.

b. The **mother star,** or **monaster,** consisting of the chromosomes radially arranged around the equator of the spindle.

c. **Metakinesis,** or the formation of two **daughter stars,** the **diaster,** a result of the longitudinal splitting of the chromosomes and the migration of the two parts towards the centrosomes.

3. The **anaphase** is characterized by : —

a. The reformation of the **nuclear network** through the fusion of the chromosomes, and the disappearance of the nuclear spindle.

b. The reappearance of the **nuclear membrane** and the **nucleolus.**

c. The division of the cytoplasm into halves.

THE TISSUES.

THE fundamental tissues are four: — Epithelial: Connective: Muscular: Nervous.

A. EPITHELIAL TISSUE.

Epithelial tissue consists of endothelium and epithelium, and is composed of a continuous layer or layers of cells covering the surface of the body and lining its cavities.

I. **Endothelium** consists of a single layer of greatly flattened cells with irregular, wavy outlines, and of widely varying shapes. It lines the abdominal and the thoracic cavities, the bursae, the blood-vessels and lymphatics, the articular cavities, and the tendon-sheaths, — *i. e.* cavities not opening directly to the exterior.

II. **Epithelium** covers the body and, in general, lines open cavities.

1. **Simple squamous epithelium** consists of a single layer of flattened cells. It is not abundant and occurs in parts of the eye, of the ear, of the ventricles of the brain, of the ducts of glands, and of the kidney. The cell-outlines are not wavy.

2. **Stratified squamous epithelium** consists of several layers of flattened cells, though the deepest layer is columnar and the middle layers often consist of **prickle-cells**, *i. e.* cells connected with one another by minute, prickle-like processes. It occurs in the whole epidermis, the cornea, the mouth, pharynx, and oesophagus, on the vocal cords, in the vagina, the female urethra, and at the beginning and end of the male urethra.

3. **Simple columnar non-ciliated epithelium** consists of a single layer of columnar cells, the free ends of which often have a cuticular secretion. It occurs in all of the digestive tract below the oesophagus, in the ducts of many glands, in the seminal vesicles, the ejaculatory ducts, and the greater part of the male urethra.

4. **Simple columnar ciliated epithelium** has a free margin of cilia to the cells instead of cuticula. It occurs in the canal of the spinal cord, in the smallest bronchi, in the oviducts and the uterus.

5. **Stratified columnar non-ciliated epithelium** consists of several layers of columnar cells, the innermost being the least elongated. It is not abundant and occurs on the inner surface of the eyelid (the palpebral conjunctiva), in the excretory ducts of some glands, and in the membranous portion of the male urethra.

6. **Stratified columnar ciliated epithelium** occurs in the Eustachian tubes, the nasal cavity, the upper part of the pharynx, the larynx, the trachea, the larger bronchi, and the epididymis.

III. Glands consist of epithelial cells.

1. Unicellular glands or **goblet-cells** occur as single, secreting cells in all columnar epithelium and are most abundant in the intestines.

2. Multicellular glands are tubular or saccular; the latter are often known as alveolar or acinous glands.

    a. Simple tubular glands are of nearly uniform calibre, branched or unbranched. Here belong the sweat glands, the small glands of the mouth-cavity, the glands of the tongue, of the stomach, of Lieberkühn, of Brunner, and of the uterus.

    b. Compound tubular glands consist of a large number of branching tubules. They include the mammary, the lachrymal, the salivary, and the large mucous glands, the thyroid, the liver, the kidneys, the prostate, the testicles and Cowper's glands.

    c. Simple saccular glands consist of branched or unbranched saccules with an excretory duct. They include the sebaceous and the Meibomian glands.

    d. Compound saccular glands consist of a number of simple saccular glands united, — *e. g.* the lungs.

B. CONNECTIVE TISSUE.

Connective tissue is characterized by a large amount of intercellular substance.

I. Mucous connective tissue possesses stellate cells with much gelatinous intercellular substance and a few fibrils. It occurs in man in the umbilical cord only, and is there known as the **jelly of Wharton**.

II. Fibrous connective tissue consists of cells and a large amount of fibrous intercellular substance imbedded in a soft homogeneous matrix. If the fibres are loosely arranged, as in subcutaneous tissue, they constitute **areolar tissue**. The fibres are white and inelastic, yielding gelatine on boiling, — with some yellowish, elastic fibres intermingled; the latter consist of elastin. The white may predominate, compactly arranged, as in tendons, constituting **fibrous tissue**; or the yellow may predominate, as in ligaments, constituting **elastic tissue**. The cells may be rounded, — *e. g.* **plasma cells**, or flattened and irregular or stellate. They may contain fat-globules, and a mass of such cells constitutes **adipose tissue**.

III. Reticular connective tissue consists of a network of fibres, — resembling white fibres, but consisting of reticulin, — containing leucocytes in its meshes. It occurs chiefly in lymph-nodules and is then known as **adenoid tissue**.

IV. Cartilage consists of a **matrix** enclosing **capsules,** which contain rounded or flattened cells.

1. **Hyaline cartilage** possesses a clear, bluish matrix, apparently structureless but containing very fine fibrils, and yields chondrin on boiling. Here belong the articular and the costal cartilages and those of the nose and the respiratory tract. In old age hyaline cartilage may become calcareous cartilage by the deposition of calcareous salts in the matrix.

2. **Elastic cartilage** is yellowish and possesses a hyaline matrix, which contains elastic fibres. It occurs in the external ear, the epiglottis, the cartilages of Wrisberg and of Santorini, and the anterior ends of the arytenoid cartilages.

3. **Fibrous cartilage** consists of a matrix largely composed of bundles of fibrous connective tissue, and of cells rather widely separated from one another. It is rare, occurring chiefly in the intervertebral discs, the symphysis pubis, and some articulations.

V. Bone consists of a matrix, — of fine fibrils and calcium salts, chiefly $CaCO_3$ and $Ca_3P_2O_8$, — and of bone-cells, which lie in lenticular cavities in the matrix, the lacunae, connected with one another by minute canals, the canaliculi. Macroscopically two kinds of bone may be distinguished, **compact bone** and **spongy bone**. Blood-vessels are present in the matrix of compact bone and lie in channels known as the **Haversian canals**. Dentine consists of bone containing no cells in its matrix but merely cell processes.

C. MUSCULAR TISSUE.

Muscular tissue consists of elongated, contractile cells of two kinds, the smooth muscle-fibres and the striated.

I. Smooth, non-striated or involuntary muscle-fibres consist of spindle-shaped, often flattened cells with tapering ends, and in each an elongated nucleus. They vary in length from 25 to 225 μ,[1] and in width from 4 to 8 or 10 μ. They occur in the skin and in the eye, in the walls of the trachea, the bronchi, the blood- and the lymph-vessels, the alimentary canal and the gall bladder, in the capsule and the pelvis of the kidney, in the walls of the ureters, the urinary bladder and the various ducts of the reproductive organs, and in erectile tissue.

II. Striated or voluntary muscle-fibres consist of greatly elongated, multinuclear cells, each enclosed in a cell membrane, the **sarcolemma**, with which the nuclei are in contact. They vary in length from about 4 to 12 cm, and in width from 12 to 60 μ. The fibres are cylindrical, with rounded or pointed ends within the muscles, but broad and blunt or notched at the tendonous attachment. They show alternating **dark** and **light transverse bands**, — the anisotropic and the isotropic. A narrow light band, the **median disc**, passes transversely through the

---

[1] 1 μ (micron) $= 0.001$ of a millimeter.

middle of the dark band, and likewise a narrow dark band, the intermediate disc, divides the light band into two, known as the lateral discs. The fibres, if hardened with alcohol, break up into longitudinal fibrils, the ultimate or contractile fibrillae, or sarcostyles, which may be divided transversely into prismatic anisotropic parts, the sarcous elements. The fibrillae are united by sarcoplasm; they extend through the entire length of the muscle-fibre. There are two kinds of striated fibres, — the dim fibres, rich in sarcoplasm, and the pale fibres, representing a higher differentiation. In man these two kinds are intermingled in the same muscle, — in some animals they occupy distinct muscles. Striated fibres occur in the muscles of the trunk and the extremities, of the ear, the eye, the tongue, the pharynx, the larynx, the upper part of the oesophagus, the diaphragm, the rectum, and the genital organs. Branched fibres occur in the muscles of the eye and the tongue.

III. Cardiac muscle-fibres are transversely striated, relatively short, and usually branched and anastomosing, and without a sarcolemma. The longitudinal fibrillae, fewer than in voluntary muscle-fibres, are radially arranged, with a mass of sarcoplasm containing the nucleus in the centre of the fibre.

D. NERVOUS TISSUE.

Nervous tissue consists of nerve- or ganglion-cells and their processes. I. Nerve-cells are distinguished as unipolar cells, with only one process, *e. g.* in the olfactory mucous membrane; bipolar cells, with two processes; and multipolar cells, with many processes. They all consist of granular or faintly striated protoplasm, containing a nucleus, but without a cell membrane. Typically each nerve-cell possesses several branched, protoplasmic processes or dendrites, — granular or faintly striated, and irregular in outline, — and one axis-cylinder process or axon, — hyaline with a smooth outline and consisting of fine fibrillae united by neuroplasm. The axon gives off occasionally fine lateral branches, the collateral fibres, and ends in a number of fine, terminal branches. A nerve-cell with such an axon is a cell of the first type. A cell whose axon divides a number of times and terminates near the cell, is a cell of the second type. A nerve-cell and its axon is known as a neuron.

II. Nerve-fibres consist of the axis-cylinder processes of nerve-cells. The axon may be naked, or surrounded by one or two sheaths, — if two, the inner or thicker is the medullary sheath, the outer or thinner is the neurilemma or Schwann's sheath. The medullary sheath consists of a hyaline, fatty substance, myelin, divided into medullary segments by oblique partitions of cement substance and entirely interrupted at intervals of from 0.08 mm to 1 mm, so that the neurilemma comes in contact with the axon, forming the nodes of Ranvier. After treatment with strong alcohol, a network appears in this sheath. The neurilemma is a delicate membrane having occasional nuclei on its inner surface. We distinguish non-medullated or gray fibres and medullated or white fibres.

1. **Non-medullated** fibres are generally naked, — *e. g.* in the olfactory nerves, some sympathetic fibres, and the peripheral terminations of many nerves.

2. **Medullated** fibres possess a medullary sheath.

a. Without a neurilemma, — found in the central nervous system only.

b. With a neurilemma, — found in the cerebro-spinal and in the sympathetic nerves.

THE ORGANS.

THE SKELETAL SYSTEM.

The skeletal system consists of bones and cartilages.

I. The bones are enclosed, except on articular surfaces, in a fibrous membrane, the periosteum, inside which is a thicker or thinner layer of compact bone. In long bones the axes are filled with the bone-marrow, which also fills the interstices of the spongy bone, which predominates in the epiphyses. The short and the flat bones consist of spongy bone enclosed in a thin layer of compact substance.

1. The periosteum consists of two layers: an outer, to which adjacent structures are attached, which contains numerous blood-vessels, and an inner, containing fewer blood-vessels, but numerous elastic fibres and connective-tissue cells ; from this inner layer fibres penetrate the matrix of the bone ; these are known as Sharpey's fibres. Cubical cells lie adjacent to the bone in places, and as osteoblasts take part in its development.

2. Compact bone exhibits a lamellar matrix due to the arrangement of its bundles of fibrillae in lamellae, which form three systems. These fibrillae are often called the decussating fibres because those in adjacent layers cross one another.

a. The Haversian system consists of an Haversian canal with 8 to 15 lamellae arranged concentrically about it. The lacunae have a similar arrangement, and the canaliculi communicate with the Haversian canal.

b. The interstitial system consists of irregularly placed lamellae, filling the interstices between the Haversian systems.

c. The circumferential or fundamental system consists of lamellae parallel with the inner and the outer surfaces of the bone. Amongst these lamellae are canals containing blood-vessels, Volkmann's canals, which are not surrounded by concentric lamellae. Canaliculi open on the inner and the outer surfaces of the bone.

Penetrating fibres similar to Sharpey's fibres may occur in the last two lamellar systems; they are attached apparently to lamellae.

3. The bone-marrow is of two kinds; the red occurs in nearly all bones except the long and the short bones of the extremities, which contain the yellow.

a. Red marrow consists of a fine connective-tissue network, containing a few fat-cells, numerous marrow-cells, which resemble leucocytes, giant-cells, which resemble huge leucocytes and are often multinuclear, and haematoblasts, which are round, nucleated cells, containing yellowish protoplasm, like colored blood-corpuscles.

b. **Yellow marrow** consists of a connective-tissue network and numerous fat-cells. Marrow-cells and haematoblasts are generally absent.

The blood-vessels pass from the periosteum into the Haversian and Volkmann's canals and thence into the marrow cavity, where they form a capillary network. Lymphatics are found in the outer layer only of the periosteum. Nerves enter the Haversian canals and the bone-marrow; in the periosteum they sometimes terminate in Pacinian bodies.

4. The **articulations** are immovable, **synarthroses,** or movable, **diarthroses.**

a. The **synarthroses** exhibit a connection by ligaments, forming a **syndesmosis,** or by cartilage, forming a **synchondrosis.** The ligaments consist of white, fibrous tissue or of elastic tissue, whose fibres are separated by loose fibrous tissue. The sutures are connected by ligaments. The cartilage is partly hyaline and partly fibrous; here belong the intervertebral ligaments.

b. The **diarthroses** consist of a ligamentous capsule and the articular cartilages. The capsule consists chiefly of elastic tissue and is lined with a **synovial membrane,** consisting of an outer portion of loose fibro-elastic tissue and fat-cells, and an inner, thin layer of fibres and connective-tissue cells, with an imperfect endothelium. The free surface of the membrane exhibits occasional fine processes, the **synovial** or **Haversian fringes** or **villi.** The capsule contains a viscid fluid, the **synovia,** consisting mainly of water. The articular surfaces consist generally of **hyaline cartilage,** — thin at the margins, — but in some cases of **fibro-cartilage** or of dense fibrous tissue. The cartilage is separated by a narrow, striated zone from a thin layer of calcareous cartilage, which directly adjoins the bone.

Blood-vessels and nerves lie in the loose connective tissue of the synovial membrane and extend into some of the villi; some of the nerves end in Pacinian bodies.

II. The **cartilages** are enclosed in a membrane, the **perichondrium,** consisting of a compact layer of fibrous and elastic tissue. This is absent on articular surfaces only. Blood-vessels and nerves occur in the perichondrium; they also occur within growing cartilage, but are absent in the adult.

THE MUSCULAR SYSTEM.

I. The **muscles** consist of striated muscle-fibres, arranged longitudinally, each of which is enclosed in a thin connective-tissue sheath, the **endomysium**; these fibres are collected into groups surrounded by connective tissue, the **perimysium**; a number of these groups constitutes a muscle, which is enclosed in the **epimysium**, a dense fibrous sheath.

II. The **tendons** and the **fasciae** consist of parallel, white fibrillae, united by cement substance into **primary bundles**, between which lie the **tendon-cells**, of various shapes. Groups of primary bundles form the **secondary bundles**, which are separated by loose fibrous and elastic tissue. Muscles and tendons are connected by the fibrous sheaths of the muscles. The fasciae sometimes contain much elastic tissue.

III. The **tendon-sheaths** and the **bursae** are essentially the same in structure as the synovial membranes.

Blood-vessels lie in the perimysium, and extremely fine capillaries penetrate the endomysium, forming networks around the fibres. They are scarce in the tendons and the fasciae, lying in the loose connective tissue between the secondary bundles. **Nerves**, both sensory and motor, supply the muscles, — see **Peripheral Nerve Endings**.

THE PERIPHERAL NERVOUS SYSTEM.

I. The **nerve trunks** consist of nerve-fibres, each surrounded by delicate connective-tissue fibres, the **fibrillar septa**; the nerve-fibres occur in groups, which are separated from one another by a fibrous sheath, the **endoneurium**, which is directly continuous with the fibrillar septa. A number of these groups constitutes a **funiculus**, enclosed in concentric lamellae of fibro-elastic tissue forming the **perineurium**; between the connective-tissue layers, **endothelioid plates** occur. When there is more than one funiculus in a nerve trunk, they are held together and enclosed in loose fibro-elastic tissue, the **epineurium**, which often contains fat-cells. Near their termination the funiculi divide until they often consist of single fibres only, enclosed in a thin fibrous sheath, which is lined with endothelioid cells and known as the **sheath of Henle.**

Blood-vessels occur in the connective-tissue membranes, and form capillary networks in the endoneurium. **Lymph-clefts** are abundant. **Nerves**, the **nervi nervorum,** occur in the epineurium.

II. The **ganglia,** usually macroscopic, consist of a **fibrous capsule,** which is a continuation of the epineurium of the nerve trunks, together with **nerve-fibres** and **ganglion-cells.** Fibrous tissue from the capsule extends into the ganglion, and surrounds the nerve-fibres and the ganglion-cells. In some ganglia, — *e. g.* that of the auditory nerve, — the cells are bipolar; in the **spinal ganglia** and several others they are bipolar in the embryo only; later the two branches fuse for a distance and then continue separately, one passing to the periphery, the other to the spinal cord and terminating in branches in the gray substance. Each cell is enclosed in a capsule of flattened connective-tissue cells, which are continuous with the neurilemma of the nerve-fibres. In this capsule, non-medullated, sympathetic fibres branch and form a plexus. In the **sympathetic ganglia** the cells are multipolar, and the protoplasmic processes surround adjacent cells.

III. The **peripheral nerve endings.**

  1. The **terminations of sensory nerves.**

    a. **Free endings** have been found in the deeper layers of the epidermis, in the cornea, in the oral mucous membrane, and between muscle-fibres. The nerve-fibre first loses its medullary sheath, then the neurilemma, and the naked axon divides and forms a plexus from which the fibrillae pass between the epithelial cells, terminating in pointed or club-shaped free ends. The significance of the branched cells of **Langerhans**, found in the epidermis, is still disputed.

b. **Tactile cells** occur in the deeper part of the epidermis and in the more superficial part of the derma. They consist of oval cells, whose deeper surface is in contact with the **tactile meniscus**, a crescentic expansion, which is connected with a non-medullated nerve-fibre. Compound tactile cells, *i. e.* two or more such cells enclosed in a capsule, occur in lower animals.

c. **End-bulbs** may be said to include tactile corpuscles, spherical end-bulbs, articular corpuscles, genital corpuscles and the Pacinian bodies.

Tactile corpuscles or **corpuscles of Meissner** or **Wagner** occur chiefly in the papillae of the derma, especially on the palmar surface of the fingers and the plantar surface of the toes. They are ellipsoid bodies, from 40 to 140 μ long and half as broad, with a fibrous capsule, and exhibit **transverse striations** produced by the elongated nuclei and the flattened boundaries of the cells. One or several medullated nerve-fibres may approach the corpuscle, and after several windings, enter it at its deeper end, lose their sheaths, and form a plexus of varicose fibrillae. The exact method of termination is still in dispute.

Spherical end-bulbs, — occurring chiefly in the conjunctiva, — and the **articular corpuscles,** — found in articulations, — have a similar structure. They are from 20 to 100 μ in diameter.

Genital corpuscles occur in the deeper part of the integument and are most numerous on the glans penis and the clitoris, — from one to four per sq. mm. They are spherical or ovoid, about 0.4 mm long, and consist of a cellular connective-tissue capsule enclosing a granular mass, in which the axis-cylinders of nerve-fibres, — entering at several points, — form a dense plexus of varicose fibrillae.

The **Pacinian bodies** or **corpuscles of Vater** occur on the nerves of the deeper parts of the integument, particularly on the hand and the foot, on the nerves of the penis and the clitoris, in the mesentery, in the periosteum and articulations of some bones and elsewhere. They are ellipsoid with a **thick capsule** composed of from 25 to 50 concentric lamellae, each consisting of two layers of connective-tissue fibres, an inner longitudinal and an outer transverse, lined with a single layer of endothelioid cells; a serous fluid separates the lamellae from one another. Along the course of the entering nerve, the lamellae are often united by a longitudinal band of tissue, the intracapsular or interlamellar ligament. The **inner bulb** occupies the centre of the capsule, and consists of a cylindrical, finely granular mass, often faintly striated, with occasional nuclei at the periphery. Through the axis of this bulb, the axon passes, terminating in a club-shaped or a slightly branched ending. A small artery enters with the nerve-fibre and sends branches to the lamellae of the capsule.

2. **The termination of motor nerves.**

a. In smooth muscle, non-medullated nerves form a network, the ground plexus, in the nodes of which ganglion-cells usually occur; from this, branches arise and form the finer intermediate plexus, and this gives off the intramuscular fibrillae, which seem to terminate as free endings on the muscle-fibres.

b. In striated muscle, medullated nerves divide a number of times, and their branches then anastomose, forming the intramuscular plexus. From this plexus small bundles of nerve-fibres arise, which divide, and generally a single fibre goes to each muscle-fibre. The nerve-fibre loses its medullary sheath on coming in contact with the muscle-fibre, and the axon divides, terminating in a number of swollen ends, which lie in a granular, nucleated mass of protoplasm, the sole-plate. The nerve-terminations and the sole-plate constitute the motor end-plate. The end-plate is said to lie under the sarcolemma, but this is disputed.

IV. The suprarenal bodies are more closely connected with the nervous system than with any other organ. Each body is enclosed in a fibrous capsule, sending delicate processes or septa into the interior, which consists of two regions, an outer, the cortex, and an inner, the medulla.

1. The cortex consists of rounded or more or less angular cells collected into groups, which are separated by fibrous tissue. These groups form three zones, — the outer or zona glomerulosa, where the groups are ovoid, the middle or zona fasciculata, where they are cylindrical, and the inner or zona reticularis, where they are irregular and anastomose, and the cells are pigmented.

2. The medulla contains polygonal cells forming cords or irregular networks, and connective tissue enclosing numerous blood-vessels, ganglion-cells and non-medullated nerve-fibres.

The arteries branch in the capsule and pass to the interior, forming capillary networks around the groups of cells of both the cortex and the medulla. In the medulla the veins unite to form the suprarenal vein; its largest branches are accompanied by longitudinal bundles of smooth muscle-fibres. The nerves, non-medullated, are very numerous, and accompany the arteries to the medulla, where they form a complex network containing ganglion-cells.

THE SKIN AND ITS APPENDAGES.

The skin consists of the epidermis and the corium.

I. The **epidermis**, cuticle, scarf-skin, or outer portion, consists typically of four layers, composed entirely of epithelial cells.

> 1. The **stratum corneum** or outermost; cells somewhat flattened, cornified, — contents dessicated.
>
> 2. The **stratum lucidum**, — wanting when epidermis is thin; cells partially cornified.
>
> 3. The **stratum granulosum**, a thin layer, cells contain eleidin or keratohyaline granules, the result of the cornification of the protoplasm.
>
> 4. The **stratum (or rete) mucosum**, or **stratum Malpighii**, the innermost. Several layers of cells, — the deepest of typical cylindrical cells covering the papillae of the corium ; the remaining layers are of polyhedral, prickle cells. This layer is more or less pigmented.

II. The **corium**, derma, true skin, or inner portion, consists of white (*i. e.* non-elastic) fibrous connective tissue and elastic fibres, cells, and smooth muscle-fibres, forming two layers blending with one another.

> 1. The **stratum papillare**, or superficial layer, consists of a dense felt-work of fibres ; the outer surface is beset with papillae.
>
> 2. The **stratum reticulare** consists of a loose network intimately connected with the stratum subcutaneum.

The **stratum subcutaneum** consists of loose fibrous tissue and adipose tissue ; the latter when dense forms the **panniculus adiposus**.

III. The **nails** consist of the thickened stratum lucidum and therefore of horny, nucleated cells, which rest on the **nail-bed**; this consists of the stratum Malpighii and the corium. The nails are separated from the **nail-walls** at the sides by the **nail-groove**. The groove at the attached end forms the **matrix**, marked by the **lunula**, and the free end projects over the **nail-ridge**.

IV. The **hair** consists of the shaft, or free portion, and the root, terminating in the hair-bulb, which surrounds a portion of the corium, the hair-papilla. The root is surrounded by the hair-follicle.

> 1. The **shaft** consists of three layers of epithelial cells.
>
> a. The **cuticle**, or outer layer, a single layer of transparent, horny, non-nucleated cells.
>
> b. The **cortical substance**, or middle layer, the thickest, consists of elongated, horny cells with attenuated nuclei; this layer is pigmented in all colored hairs except white.
>
> c. The **medulla**, or central portion, — wanting in many hairs, — consists of rows (usually two) of irregular cuboidal cells.

2. The **root** and **hair-follicle** below the sebaceous glands consists of four regions :

a. The **fibrous-sheath,** or outermost, consists of three layers derived from the corium :

The **longitudinal fibrous layer,** the outermost.

The **circular fibrous layer.**

The **glassy membrane.**

b. The outer **root-sheath,** a continuation of the rete mucosum.

c. The **inner root-sheath** consists of two layers : —

Henle's layer, the outer, of non-nucleated cells in one or two layers ;

Huxley's layer, the inner, of nucleated cells in a single layer.

d. The **root** consists of the same layers as the shaft.

V. The **glands** of the skin.

1. **Sebaceous glands** occur in the stratum papillare in connection with all hairs and also on the edge of the lips, on the eyelids, the labia minora, the glans, and the prepuce, and are simple, saccular glands, branched or unbranched. Those connected with hairs lie between the arrector pili muscle and the hair-follicle and open into the follicle. The duct is lined with stratified, squamous epithelium, — the sacs with cuboidal. The secretion is sebum.

2. **Sweat glands** occur over almost the entire skin and consist of the coil or secreting portion and the long, wavy, excretory duct.

a. The **coil** is lined with a single layer of columnar epithelium resting on a membrana propria, sometimes with smooth muscle-fibres between the cells and the membrane.

b. The **duct** is lined with two layers of low, cuboidal cells and surrounded by longitudinal bundles of fibrous connective tissue.

VI. **Blood-vessels, lymphatics,** and **nerves** of the skin.

1. The large **arteries** lie beneath the skin and send up branches from which three sets of vessels arise, supplying, —

a. The subcutaneous adipose tissue.

b. The sweat glands.

c. The papillae, hair-follicles, sebaceous glands, etc.

2. The **veins** follow in general the course of the arteries.

3. The **lymphatics** form two sets, —

a. In the corium with special networks around the hair-follicles and the glands.

b. In the subcutaneous layer.

4. The large **nerves** lie in the subcutaneous tissue and send branches superficially, which form the **subpapillary plexus,** whose fibres end in tactile cells, tactile-corpuscles, or intra-epithelial fibrils. The subcutaneous tissue contains Pacinian bodies.

THE CIRCULATORY SYSTEM.

A. The Blood System.

I. The heart consists of three layers : —

1. The **endocardium**, or innermost, consists of, —

a. An endothelial lining.

b. A fibrous connective-tissue layer containing elastic fibres and smooth muscle-fibres.

c. An outer connective-tissue layer continuous with the perimysium of the muscular layer.

2. The **muscular layer** consists of a network of naked, branched, striated muscle-fibres, arranged in two layers in the auricles, — an inner longitudinal and an outer transverse; in the ventricles they are placed more irregularly. The spaces between the muscle-fibres are lymph spaces.

3. The **pericardium** is separated from the muscular layer by adipose tissue and consists of, —

a. A fibrous connective-tissue layer containing elastic fibres.

b. An external endothelial layer.

4. The **valves** consist of fibrous connective tissue covered with endothelium.

II. The arteries are small, medium, or large, and their walls consist of three coats, — the **intima**, or inner, the **media**, and the **adventitia**, or outer.

The **small arteries** are the terminal branches near the capillaries.

1. The **intima** consists of two layers : —

a. An **endothelium** lining the vessels and consisting of elongated cells.

b. The **internal elastic membrane**, structureless.

2. The **media** consists of a single, circular layer of smooth muscle-fibres.

3. The **adventitia** consists of longitudinal bundles of fibrous connective tissue and elastic fibres.

The **medium arteries** are all except the aorta and the pulmonary artery.

1. The **intima** consists of three layers : —

a. The **endothelium**.

b. The **subendothelial layer** consists of fibrous connective tissue, flattened, branched cells, and elastic fibres.

c. The **internal elastic membrane**.

2. The **media** consists of several, circular layers of smooth muscle-fibres and networks of elastic tissue. Sometimes longitudinal muscle-fibres are also present.

3. The **adventitia** consists of the **external elastic membrane,** which adjoins the media, and fibrous connective tissue with scattered, longitudinal, smooth muscle-fibres, and connective-tissue cells.

The **large arteries** are the aorta and the pulmonary.

1. The **intima** consists of three layers : —
   a. The **endothelium,** of broad, polygonal cells.
   b. The **subendothelium,** of fibrous connective tissue, cells, and a network of elastic fibres.
   c. The **fenestrated membrane of Henle,** which is a modification of the internal elastic membrane.

2. The **media** consists of circular layers of smooth muscle-fibres alternating with networks of broad elastic fibres.

3. The **adventitia** is as in medium-sized arteries, but without the external elastic membrane.

III. The **veins** show essentially the same structure as the arteries, but fibrous connective tissue predominates over elastic fibres and muscular tissue except in the adventitia, where the latter is more abundant than in the arteries.

IV. The **capillaries** connect the arteries and the veins with a few exceptions, — *e. g.* the spleen, corpora cavernosa, tips of fingers, etc. The walls of capillaries contain no muscle-fibres and consist of a single layer of endothelium only.

V. The **blood-vessels, lymphatics,** and **nerves** of the arteries and the veins occur in the large and the medium-sized vessels, and lie almost exclusively in the adventitia and never in the intima.

VI. The **blood** consists of a nearly colourless fluid, the **blood-plasma,** and cellular elements, the **blood-corpuscles,** coloured or colourless, together with **blood-platelets.**

1. The **coloured blood-corpuscles** are non-nucleated cells consisting of a colourless **stroma,** which contains the colouring matter, or **haemoglobin.** They are greenish-yellow, biconcave discs with rounded edges, about 7.5 μ in diameter and 1.5 μ thick.

2. The **colourless blood-cells,** or **leucocytes,** occur in the blood, the lymph, and between the cells of various tissues. They are amoeboid, with a nucleus and without a cell membrane and average about 10 μ in diameter. There is about one white corpuscle to 300 or 400 coloured corpuscles in the blood.

3. The **blood-platelets** are colourless discs, circular or oval, about one-third the diameter of the coloured corpuscles and occurring in varying numbers.

VII. The **arterial glands,** the carotid and the **coccygeal** or **Luschka's** gland, are not glands. They consist of dense networks of small blood-vessels surrounded by groups of rounded cells, fibrous tissue, nerve-fibres and ganglion-cells.

B. The Lymphatic System.

I. Lymph spaces occur as clefts throughout the connective tissue of the body, often with an imperfect, endothelioid lining, and connect with one another and with the lymph capillaries. The large serous cavities, *e. g.* peritoneal and pleural, are enlarged lymph spaces; they possess minute openings, the **stomata.**

II. The **lymph capillaries** possess walls of a single layer of endo-thelial cells and show frequent constrictions and dilatations.

III. The **lymph vessels** have walls composed of three layers: —

1. The **intima,** or inner, consists of endothelium and elastic fibres. In the thoracic duct fibrous tissue is also present.

2. The **media** consists of circularly arranged smooth muscle-fibres with a few elastic fibres.

3. The **adventitia** is composed of fibrous connective tissue.

IV. The **lymph** consists of a colourless fluid and cells, the **lymph-corpuscles** or leucocytes, together with fatty granules, which occur mainly in the lacteals of the intestine.

V. Lymphoid, or **adenoid tissue** consists of a connective-tissue net-work and round cells or leucocytes.

1. Diffuse adenoid tissue occurs in undefined masses in many mu-cous membranes, *e. g.* of the respiratory and the digestive tracts.

2. Simple lymphatic nodules, or **solitary follicles,** occur in many mucous membranes in the tunica propria, and consist of adenoid tissue surrounded by a fibrous connective-tissue capsule.

3. The **compound lymphatic follicles,** or **lymph-nodes** (incorrectly called " glands "), occur on the lymph-vessels, which as afferent ves-sels enter the follicles at various points, divide and form networks, the **lymph-sinuses,** and reuniting pass out at the **hilum** as efferent vessels. A capsule of fibrous connective tissue surrounds the folli-cles and sends rods, or trabeculae, into the interior. The follicles show two regions: —

a. The **cortex** consists of simple follicles, separated by lymph-sinuses and trabeculae.

b. The **medulla,** continuous with the cortex, consists of anasto-mosing rods of adenoid tissue, surrounded by lymph-sinuses.

Small blood-vessels enter the follicles at various points and supply the capsule and the trabeculae. A large artery enters at the **hilum** and supplies the adenoid tissue. The veins pass out at the hilum. Nerves are few.

4. The **spleen** is a compound lymphatic follicle formed of a capsule of fibrous connective tissue and elastic fibres, from which trabeculae pass into the interior and form a loose network, in the meshes of which lies the **splenic pulp,** which consists of blood-vessels and adenoid tissue. At various points the arteries are surrounded by spherical or ovoid masses of dense adenoid tissue precisely like the simple lymphatic nodules ; these are called **Malpighian corpuscles.** The **arterial capillaries** open into the **venous capillaries,** which consist of spaces in the splenic pulp, containing **leucocytes, coloured blood-corpuscles,** and **pigment granules.** The large blood-vessels pass through the **hilum. Lymphatics** are mainly confined to the capsule and the trabeculae, and the **nerves** follow in general the course of the arteries.

5. The **thymus body,** of epithelial origin, becomes a lymph-follicle by an ingrowth of connective tissue, and later the adenoid tissue is replaced by adipose. It resembles a compound lymphatic follicle except that the cortex is denser than the medulla, and that it contains oval bodies with concentric markings, the **corpuscles of Hassall,** — the remains of the original epithelial structures.

THE RESPIRATORY SYSTEM.

I. The **larynx** consists of a cartilaginous framework surrounded by fibrous and by muscular tissue and lined with a **mucous membrane** consisting of an **epithelium**, a **tunica propria** and a **submucosa**.

1. The **epithelium** is **ciliated stratified columnar** except on the epiglottis, on the region between the epiglottis and the false vocal cords, and on the true vocal cords, where it is **stratified squamous**. Taste-buds occur on the posterior surface of the epiglottis.

2. The **tunica propria** consists of white and of elastic fibrous connective tissue. In the true vocal cords it consists almost entirely of elastic tissue. Beneath the squamous epithelium it forms papillae. Numerous leucocytes are present.

3. The **submucosa** consists of loose connective tissue and contains branched tubular glands from 0.2 mm to 1 mm in length. **Blood-vessels** and **lymphatics** are numerous in the subepithelial tissue, forming two or three networks parallel to the surface. The **nerves** contain microscopic ganglia and often terminate in end-bulbs.

The thyroid, the cricoid, and the greater part of the arytenoid **cartilages** are hyaline. The median part of the thyroid, the apex and vocal processes of the arytenoids, the cartilages of Wrisberg and of Santorini and of the epiglottis are yellow elastic.

II. The **trachea** is like the lower portion of the larynx except that the elastic fibres are more numerous and arranged longitudinally, and that smooth muscle-fibres extend transversely between the ends of the imperfect, cartilaginous rings. The mucous glands of the dorsal wall are often 2 mm in length.

III. The **bronchi** by branching become continually smaller until they have a diameter of 0.5 mm, when they become continuous with the **terminal bronchi** or **bronchioles**, which in turn open into the **alveolar ducts**, leading to the **terminal vesicles**, or **infundibula**.

1. The larger **bronchi** have the same structure as the trachea; in the smaller the cartilaginous rings are replaced by irregular **cartilaginous plates**, which disappear entirely in the smallest bronchi, while the smooth muscle-fibres form a complete ring or **muscularis mucosae**, and extend as far as the infundibula. The **mucous glands** extend as far as the terminal bronchioles.

2. The **terminal bronchioles** contain a few evaginations or **alveoli** on their walls. There is at first a **simple columnar ciliated epithelium**, succeeded by a **non-ciliated cuboidal**, and this in turn by the **respiratory epithelium**, adjoining the alveolar ducts, which con-

sists of cuboidal cells mingled with flattened, non-nucleated plates. Outside the epithelium are longitudinal **elastic-fibres** and irregular, **smooth muscle-fibres** circularly arranged.

3. The **alveolar ducts** contain a larger number of **alveoli** than the bronchioles, and are lined with **respiratory epithelium** resting on a thin layer of **fibro-elastic tissue** and **smooth muscle-fibres**.

4. The **infundibula** consist of **alveoli** lined with **respiratory epithelium.** Muscle fibres are wanting, but numbers of **elastic fibres** are present.

IV. The **lungs** correspond to compound, saccular glands. The respiratory portion is divided into lobules, 0.3 mm to 3 mm in diameter, by areolar tissue, — the **interlobular connective tissue,** — which in the adult contains more or less **pigment.** The surface of the lungs is covered by the **visceral pleura,** consisting of fibrous and elastic tissue covered with endothelium. The **parietal pleura** is thicker and contains less elastic tissue. **Stomata** occur in the endothelium and connect indirectly with lymph-spaces. The **arteries** enter at the hilum of the lungs and follow the course of the bronchi and their divisions, ending in capillary networks beneath the respiratory epithelium; there the veins arise and pass out along the same course. From the bronchial arteries, branches pass to the walls of the bronchi, forming a deep plexus around the muscles and the glands and a superficial plexus beneath the epithelium. **Lymphatics** arise in the subpleural and in the interlobular connective tissue, follow the bronchi, receiving branches which arise in the subepithelial layers of their walls, and pass out at the hilum to the **bronchial lymph-nodules.** The **nerves,** from both the vagus and the sympathetic, both medullated and non-medullated, follow in general the course of the blood-vessels.

V. The **thyroid gland** is a compound tubular gland consisting of a large number of **closed tubules** of different sizes, whose embryonic, excretory duct, the thyro-glossal, has become closed. They are lined with **cuboidal epithelium** and contain a **colloid substance.** The tubules are surrounded by loose connective tissue, which groups them into **lobules.** The blood-vessels, lymphatics, and nerves are very numerous and lie close to the epithelium of the tubules.

THE DIGESTIVE SYSTEM.

All cavities communicating with the outside are lined with a **mucous membrane** or **mucosa**, which consists of **epithelium** resting on a connective-tissue stroma, the **tunica propria**, made up of fibrous and elastic tissue and cells, with sometimes in its deeper portion, non-striated muscle-fibres forming the **muscularis mucosae.** The surface of the tunica propria adjoining the epithelium is often modified into a **basement membrane**, or **membrana propria**, which may be structureless or composed of flattened, connective-tissue cells. Beneath the tunica propria is the loose connective tissue of the **submucosa**.

I. The **mouth.**

1. The **mucous membrane** consists of **stratified squamous epithelium** resting on a typical **tunica propria**, whose surface is beset with numerous papillae. The **submucosa** is loose except on the gums and the hard palate, where it is firmly united to the periosteum. **Mucous glands** are numerous except on the gums and the hard palate, and according to their position are known as labial, buccal, lingual, and palatine. They are branched tubular glands, from 1 to 5 mm long, lying in the submucosa, and secreting **mucin.** The excretory ducts are lined with stratified squamous epithelium, — the ends of the tubules with simple columnar epithelium, between which and the basement membrane, crescentic groups of cells often occur, the **crescents of Giannuzzi** or the **demilunes of Heidenhain,** which may represent exhausted cells. The **blood-vessels** lie in the submucosa and send branches to the tunica propria, forming a sub-epithelial, capillary network. The **lymphatics** have a similar arrangement. The **nerves** form a network in the submucosa and send branches to the tunica propria, terminating in **end-bulbs** or in **free endings** between the epithelial cells.

2. The **teeth** consist of the **crown** or exposed portion, and the **fang,** — the two parts joining at the **neck.** A tooth is composed of an inner **pulp-cavity,** enclosed on all sides except at the deeper end of the fang by **dentine,** which in turn is covered on the crown by **enamel,** and on the fang by **cementum.**

a. The **enamel,** the hardest part of the tooth, consists of parallel **prisms,** mostly hexagonal, united by a small amount of cement substance, and arranged in general at right angles to the surface of the tooth. Their outlines are slightly wavy. When first developed, a thin, resistant membrane, **Nasmyth's membrane,** extends over the outer surface of the enamel; later this wears away.

b. The **dentine**, or ivory, harder than bone, consists of a matrix of fine fibrillae and calcareous salts, through which pass radial canals, the **dentinal tubules**, — at their origin from 2.5 to 5.5 μ in diameter, — from which numerous lateral branches are given off. The denser portion of the matrix, which surrounds the tubules, forms the **dentinal sheaths**. Near the periphery are the **interglobular spaces** into which many tubules open. The projections of the dentine into these spaces are known as the **dentinal globules**. In the tubules lie protoplasmic processes from the cells of the pulp-cavity, forming the **dentinal fibres**.

c. The **cementum** has the same structure as bone. Haversian canals are generally absent. Sharpey's fibres are numerous.

d. The **pulp** consists of soft, connective-tissue cells, stellate or spindle shaped, but, adjoining the dentine, they are arranged as a compact layer of elongated cells, the **odontoblasts**, which send processes into the dentinal tubules. **Blood-vessels and nerves** are present in the pulp.

3. The **tongue** consists mainly of **striated muscles** covered with a **mucous membrane**.

The **muscle-bundles** lie in three planes: **vertical** and slightly **radial** (geniohyoglossus, lingualis, and hyoglossus); **transverse** (lingualis); **longitudinal** (lingualis and styloglossus). The **septum linguale**, a longitudinal, vertical membrane of compact connective tissue, divides the tongue into halves.

The **mucous membrane** is thickest on the superior surface, where it is characterised by the papillae, of which there are three kinds, — the **filiform** or **conical**, the **fungiform**, and the **circumvallate**.

a. The **filiform papillae** occur over the whole superior surface of the tongue, and consist of conical elevations of the tunica propria, bearing small secondary papillae, and covered with stratified squamous epithelium, which is roughened and cornified by the abrasion of the superficial cells.

b. The **fungiform papillae** are fewer than the conical but have a similar distribution. They are rounded elevations with secondary papillae on the tunica propria; the epithelium is not cornified.

c. The **circumvallate papillae** occur at the posterior end of the tongue; there are usually 8 or 10 only. They are broader and lower than the fungiform, and surrounded by a circular **groove** bounded by a **wall**. Secondary papillae occur on the tunica propria on the superior surface only. Numerous **taste-buds**, extending the whole thickness of the epithelium, occur on the sides of the papillae in the groove. These consist of a capsule of elongated cortical-cells, with a minute opening, the **taste-pore,** and enclosing the **gustatory-cells,** which often terminate peripherally in stiff, hair-like processes. Nerve-fibres terminate as free-endings between the gustatory-cells. **Taste-buds** also occur on other parts of the tongue and on the folds or **papillae foliatae,** which lie on either side of the tongue.

d. The glands of the tongue are either **mucous** or **serous.**

The **mucous glands** occur at the edges and the root of the tongue and are like those of the oral mucous membrane; the ducts are lined with columnar epithelium, often ciliated, and no demilunes are present in the tubules.

The **serous glands** occur near the circumvallate and the foliate papillae only, and have a thin, watery secretion, rich in albumen.

e. Numerous **lymph follicles** occur in the mucous membrane between the circumvallate papillae and the epiglottis, where they appear externally as circular elevations, 0.1 to 4 or 5 mm in diameter, with a minute opening leading into a **tubular crypt,** about which are lymph nodules surrounded by loose adenoid tissue and the whole enclosed in a fibrous sheath. The epithelium lining the crypt is infiltrated with leucocytes, which pass thus into the mouth.

f. The **blood-vessels** form a network in the mucosa, from which branches arise, ending in capillary networks in the papillae and around the glands. Similar networks occur in the lymph follicles and around the muscle-fibres.

The **lymphatics** form a deep set and a superficial set, — the latter receiving the lymph-vessels from the papillae. Networks surround the nodules at the root of the tongue.

The **nerves,** the glosso-pharyngeal and the lingual branch of the fifth, supply the mucous membrane, ending as elsewhere in the mouth or in close connection with the taste-buds.

4. The **tonsils** consist of numerous **lymph nodules** surrounded by loose adenoid tissue and a fibrous capsule, precisely like those at the root of the tongue. They are the chief source of the **salivary corpuscles,** found in the saliva, which, as leucocytes, migrate through the epithelium.

5. The **salivary glands** are compound tubular glands; there are three pairs opening into the mouth: —

a. The **parotid,** a **serous** gland, is divided into lobes and lobules by fibrous connective tissue containing leucocytes. The excretory duct, or **duct of Stenson,** consists of fibro-elastic tissue with a compact membrana propria lined with a double layer of columnar epithelium, — the cells in the layer adjoining the basement membrane being the smaller; this duct divides and becomes continuous with the **intralobular ducts**, lined with a single layer of columnar cells often known as **rod epithelium** because of the presence of vertical striations at the base of the cells; these pass into the **intermediate tubes** or **intercalated tubules**, lined with flattened cells, and thus connect with the **terminal compartments**, which are lined with the cuboidal secreting cells. These cells when resting appear granular and small, — when active, clearer and larger.

b. The **sublingual**, a **mucous** gland, is essentially the same in structure as the parotid, but the radial striations in the epithelium are less abundant and the intercalated tubules are apparently absent; the demilunes of Heidenhain are large. In the excretory duct, the **duct of Bartholin**, the membrana propria is not as marked as in Stenson's duct.

c. The **submaxillary**, a **mixed** gland, *i. e.* having both mucous and serous secretions, is like the parotid in structure, but the rod epithelium is more abundant, and some of the terminal compartments are lined with serous gland-cells, others with mucous cells and demilunes. The excretory duct, or **duct of Wharton**, is like Stenson's duct, but has in addition a longitudinal layer of muscle-fibres outside the connective tissue.

The **arteries** follow the ducts, terminating in capillary networks around the gland-cells. The **veins** have in general the same course. Little is known concerning the **lymphatics**.

The **nerves** are numerous, both medullated and non-medullated, and contain numerous, microscopic ganglia. Medullated fibres form networks around the tubules.

II. The **pharynx** is lined with a mucosa, outside which is a fibrous coat, a muscular and then a second fibrous coat.

1. The **mucosa** possesses **stratified ciliated columnar epithelium** in the **respiratory part** of the pharynx; here the adenoid tissue is very abundant, especially between the openings of the Eustachian tubes where it forms the **pharyngeal tonsil**, similar in structure to the palatine tonsils. In the **digestive part** of the pharynx the epithelium is **stratified squamous**, and the tunica propria has numerous papillae and mucous glands. The **sub-mucosa** unites the mucosa with the fibrous coat.

2. The **fibrous coat** or **pharyngeal aponeurosis** is a compact felt-work of fibro-elastic tissue.

3. The **muscular coat** lies between the two fibrous coats and consists of striated muscle-fibres.

4. The **outer fibrous coat** consists of a feltwork of fibro-elastic tissue.

The blood-vessels, lymphatics, and nerves are arranged in general as in the oral mucosa.

III. The **oesophagus** consists of a mucous, a muscular, and a fibrous coat.

1. The **mucosa** is like that of the lower portion of the pharynx, except that longitudinal, involuntary muscle-fibres occur in the tunica propria, forming a **muscularis mucosae,** separated from the muscular coat by a loose sub-mucosa.

2. The **muscular coat** consists of striated fibres in the upper portion of the tube, — of non-striated in the lower portion, where they are arranged in two layers, an inner circular and an outer longitudinal.

3. The **fibrous coat** consists of fibrous and elastic tissue.

Blood-vessels, lymphatics, and **nerves** are arranged as in the pharynx.

IV. The **stomach** consists of a mucosa, a submucosa, a muscular coat, and a serous or fibrous coat.

1. The **mucosa** consists of an epithelium, a tunica propria, and a muscularis mucosae.

The **epithelium** is simple columnar; the nuclei are at the bases of the cells and the outer portion gives rise to a mucoid secretion; the cells often resemble goblet cells. The **tunica propria** consists of fibrous and reticular tissue containing occasional, simple, lymphatic nodules. The **muscularis mucosae** consists of several layers of smooth muscle-fibres lying in various directions and occasionally between the tubules of the glands.

The **glands** in the cardiac and the middle thirds of the stomach are the **fundus** or **peptic glands,** in the pyloric third, the **pyloric glands.** They are all simple tubular glands, extending the whole thickness of the tunica propria, and consist of a **duct** with a pit-like opening on the surface of the mucosa, a **neck,** and a **body,** which is sometimes divided and usually tortuous. In the **fundus glands** the body is lined with the columnar, **chief** or **central cells** intermingled with the less numerous, deeper lying, **parietal** or **acid cells.** The latter are the more granular and are surrounded by a net-work of canaliculi, which open into a lateral canal lying between the central cells and thus communicate with the lumen of the gland. The **pyloric glands** are lined with columnar epithelium throughout.

2. The **submucosa** consists of loose fibro-elastic tissue, and sometimes contains fat-cells. Both the mucosa and the submucosa take part in the formation of the **rugae.**

3. The **muscular coat** consists in the pyloric region of an inner circular and an outer longitudinal layer of smooth muscle-fibres, which in the cardiac region are arranged in several directions.

4. The **serous coat** consists of fibro-elastic tissue covered on the outer surface with endothelium.

The **arteries** give off branches to the serous and the muscular coats and form a network in the submucosa parallel to the surface; from this, branches pass through the muscularis mucosae and form a second network in the deeper part of the tunica propria, from which capillaries arise, forming plexuses around the glands and beneath the epithelium. The capillaries connect with the **veins**, which in general follow the course of the arteries.

The **lymphatics** arise as blind capillaries just beneath the epithelium, pass between the glands and form a plexus in the deep part of the tunica propria. This connects with a plexus in the submucosa, from which vessels pass through the muscular coat, receiving branches from the intramuscular plexus.

The **nerves** form a plexus between the layers of the muscular coat, **Auerbach's plexus**, which contains numerous microscopic ganglia, and from this branches go to the serous coat, to the muscles, — terminating on their surfaces, — and to the submucosa, where a second, more delicate plexus is formed, **Meissner's plexus**, from which branches go to the tunica propria, terminating apparently beneath the epithelium.

V. The **intestines** consist of a mucosa, a submucosa, a muscular and a fibrous coat.

1. The **mucosa** of the small intestine is thrown into folds, the **valvulae conniventes**, mostly transverse and most numerous in the duodenum and the jejunum; it also possesses very numerous elevations, more or less cylindrical, about 1 mm high, the **villi**. Minute depressions occur in the whole intestine, — the **follicles, crypts, or glands of Lieberkühn**, which are lined with simple columnar epithelium; they extend the whole depth of the mucosa, and in the small intestine open between the villi. They are deepest in the large intestine. The **mucosa** consists of epithelium, tunica propria, and muscularis mucosae.

a. The **epithelium** is everywhere simple columnar, resting on a basement membrane. Its free surface is covered with a cuticular layer, the **basilar border**. **Goblet cells** are numerous, and migratory **leucocytes** often occur between the epithelial cells.

b. The **tunica propria** consists of fibrous and of reticular tissue lying at the base of and between the crypts, and extending into the villi.

c. The **muscularis mucosae** is made up of smooth muscle-fibres, — an inner circular and an outer longitudinal layer. Fibres also extend into the villi.

2. The **submucosa** is of loose fibrous tissue, containing in the

upper part of the duodenum the glands of Brunner, branched, tubular, serous glands lined throughout with simple columnar epithelium.

Simple lymphatic nodules, from 0.5 to 2 mm in diameter occur throughout the intestine, lying at first in the tunica propria and later extending into the submucosa. They are frequently pear-shaped, and the narrow end forms a projection on the epithelial surface.

Peyer's patches, of which from 20 to 30 are generally present, are groups of simple lymphatic nodules, some 10 to 60, surrounded by diffuse adenoid tissue and occurring in the lower two-thirds of the small intestine. They vary from 12 to 120 mm in length, and from 12 to 25 mm in width. Lymphatic follicles are usually very abundant in the vermiform appendix.

3. The muscular coat consists of two layers of smooth muscle-fibres, — a thick, inner, circular layer, and a thinner, outer, longitudinal layer. In the large intestine the saccules are produced by the longitudinal muscle-layer, which is shorter than the other layers and chiefly confined to three bands, 10 to 15 mm broad.

4. The serous coat is like that of the stomach.

The blood-vessels are arranged precisely as in the stomach, except in the small intestine, where special arteries pass into the villi and break up into capillary networks. Similar networks surround the glands of Brunner and the lymphatic nodules.

The lymph-vessels of the large intestine are arranged as in the stomach. In the small intestine they arise in the villi as canals, closed at the inner end, — the lymph-radicles or lacteals. These join the network at the base of the crypts and then continue as in the stomach. Lymph-vessels are absent in Peyer's patches.

The nerves are distributed as in the stomach.

VI. The pancreas is a serous, salivary gland. The excretory duct, the duct of Wirsung and Santorini, is lined with simple columnar epithelium surrounded by fibrous tissue, and is directly continuous with the intermediate tubules, which are lined with flattened cells, and end in the terminal compartments, which have a low columnar or conical epithelium. Intralobular tubes are therefore absent. The secreting cells contain numerous, highly refractive granules, the zymogen granules, which lie near the lumen of the tubule; the nucleus lies in the deeper, clear portion of the cell. Amongst the tubules occur lighter groups of cells, the bodies of Langerhans, which probably represent immature tubules. The blood-vessels, lymphatics, and nerves are arranged as in the other salivary glands.

VII. The liver is a compound, tubular gland, enclosed in a capsule consisting of a serous coat and of fibro-elastic tissue, which sends fibres into the interior dividing it into more or less distinct lobules, which are prismatic, about 2 mm high and 1 mm broad.   This inter-lobular connective tissue is known as the capsule of Glisson and con-tains the branches of the excretory or hepatic duct, of the portal vein, of the hepatic artery, of the lymphatics, and of the nerves.

1.   The lobules consist in part of radially arranged, often branching rods of hepatic cells without a cell membrane, amongst which are delicate fibres from the interlobular connective tissue, forming a fine network.

2.   The hepatic duct and its larger branches are lined with simple columnar epithelium containing a few goblet cells and surrounded by a tunica propria of fibrous tissue containing short mucous glands and a few smooth muscle-fibres arranged longitudinally and trans-versely.   Outside the tunica propria is a submucosa.

3.   The structure of the cystic duct, the ductus choledochus, and the gall-bladder is essentially like that of the hepatic duct; the tunica propria possesses numerous anastomosing elevations or rugae, and in addition there is a layer of interlacing, smooth muscle-fibres. The epithelial cells in the gall-bladder are about twice as high as those in the ductus choledochus.

4.   The hepatic duct branches, forming the interlobular bile-ducts, whose walls become thinner until they consist of low, epithelial cells only.   On entering the lobules, the epithelium disappears, and the ducts become minute, tubular spaces between the hepatic cells, — the bile-capillaries.   As a rule several of these lie in contact with each cell, and each tubule is surrounded by two cells instead of by several as in other glands.

5.   The vasa aberrantia are isolated, blind bile-ducts, — embryonic remains, — which occur at the left border of the liver and about the vena cava and the portal fissure.

6.   The portal vein is homologous with the artery in other glands. It enters at the transverse fissure, divides in the interlobular con-nective tissue forming the interlobular veins, from which capil-laries pass into the lobules, anastomose freely, and connect with the central or intralobular vein, which in turn opens into the sub-lobular vein at the base of the lobule.   The sublobular veins open into the hepatic vein.   The venous capillaries in the lobules usually lie at the angles of the hepatic cells, — the bile-capillaries between their flattened surfaces.

7.   The hepatic artery has in general the course of the portal vein, but its branches terminate in capillary networks in the interlobular connective tissue and in the capsule, connecting with the branches of the portal vein, and do not enter the lobules.

8.   The lymphatics follow the branchings of the portal vein, pass

into the lobules (this is a moot point), and also form a network in the capsule.

9. The **nerves**, mostly non-medullated, follow the course of the hepatic artery and contain ganglia. Their mode of termination is unknown.

VIII. The **peritoneum** is made up of bundles of fibrous and elastic tissue covered with endothelium and connected with underlying structures by loose fibrous and elastic tissue, the **subserous layer**. Stomata occur in the endothelium and connect with the lymphatics. Bloodvessels are few. The nerves, not numerous, often end in Pacinian bodies.

THE EXCRETORY SYSTEM.

I. The **kidneys** are compound tubular glands. **Fibrous tissue** encloses each kidney as a thin layer, — which also contains smooth muscle-fibres, — and occurs within the organ in a small amount surrounding the tubules and the blood-vessels. Macroscopically the kidney exhibits two regions, the **cortex** or outer third, in which the tubules and the blood-vessels are tortuous, and the **medulla** or inner two-thirds, in which they are straight. Groups of straight tubules extend from the medulla into the cortex constituting the **medullary rays** or **pyramids of Ferrein.** The medulla is divided into the **Malpighian pyramids,** each terminating in a papilla, and separated from one another by the **columns of Bertini,** which contain the larger blood-vessels. The **tubules,** beginning with the closed end in the cortex and ending with the opening at the tip of the papilla, consist of the following parts, each of which is lined by a single layer of epithelial cells, resting on a basement membrane:

1. The **Malpighian corpuscles** consist of two parts: —

a. The **capsule of Bowman** is the spherically enlarged end of the tubule, which is invaginated, forming a cup, whose walls consist of flattened cells.

b. The **glomerulus** is a mass of capillary blood-vessels, which lie in the invaginated part of the capsule.

2. The **neck** marks the abrupt transition from the enlarged capsule to the narrow uriniferous tubule.

3. The **proximal convoluted tubule** has a tortuous course extending from the neck through the cortex as far as the medulla, and is lined with low columnar rod-epithelium.

4. The **descending limb of Henle's loop** is straight, lying within the medulla, and is lined with flattened cells having projecting nuclei.

5. Henle's loop connects the descending with the **ascending limb;** the latter is also straight, extending into the cortex in a medullary ray. Both the loop and the ascending limb are lined with cuboidal epithelium.

6. The **intercalated tubule** is continuous with the ascending limb and has a tortuous course in the cortex; it exhibits two regions, — the **irregular tubule** and the **distal convoluted tubule.** It is lined with columnar rod-epithelium like the proximal convoluted tubule.

7. The **collecting tubules** lie in the medullary rays and in the medulla, increasing in size as they approach the papilla.

An arched collecting tubule connects the convoluted tubule with the straight collecting tubule. They are lined with columnar epithelium. The largest of the collecting tubules are known as the tubes or ducts of Bellini.

The blood-vessels. The branches of the renal artery pass through the columns of Bertini as far as the outer part of the medulla, where they form arches parallel with the outer surface of the kidney. From these arches branches pass to the cortex, known as the interlobular arteries, and these in turn send branches, the afferent arteries, to the glomeruli. Each glomerulus has a single afferent artery, which divides into capillaries in the glomerulus and emerges as the efferent artery, which then forms a capillary network around the tubules, with rounded meshes about the convoluted tubules and elongated meshes in the medullary rays. This network connects with the interlobular veins, which have the same course as the interlobular arteries, and connect with venous arches between the cortex and the medulla. At the periphery some of the branches of the network unite to form the venae stellatae before joining the interlobular veins. The arteries of the medulla, the arteriae rectae, arise chiefly as branches of the arterial arches and of the deeper efferent arteries, and form capillary networks about the collecting tubules. These networks connect with the venae rectae, which follow the arteries.

Lymphatics occur at the periphery of the kidney and accompany the blood-vessels.

Nerves accompany the arteries; their ultimate termination is unknown.

II. The ureters together with the calices and the pelvis of the kidney consist of three coats.

1. The mucous membrane consists of stratified squamous epithelium having only a few layers of cells and therefore sometimes known as transitional epithelium, a tunica propria of fibro-elastic tissue and connective-tissue cells, and a loose submucosa.

2. The muscular coat consists of two layers of smooth muscle fibres, an inner longitudinal and an outer circular; the lower part of the ureter has in addition an outer longitudinal layer. Around the papillae of the kidney the circular muscles form a sphincter.

3. The fibrous coat consists of loose connective tissue.

Blood-vessels and lymphatics are numerous in the mucosa.

Nerves occur in the muscular coat and in the tunica propria.

III. The urinary bladder exhibits the same structures as the ureters except that the muscular coats consist of three layers, an inner longitudinal, a middle circular, and an outer longitudinal, — the circular layer forms the internal vesical sphincter at the base of the bladder. The nerves show microscopic groups of ganglia.

IV. The **urethra** consists of a mucous membrane surrounded by a muscular coat and fibrous tissue.

1. The **female urethra** usually possesses a **stratified squamous epithelium** resting on a basement membrane, which covers the papillae of the tunica propria. Near the meatus small groups of simple, branched, tubular glands occur, the **periurethral glands.** The **muscular coat** is composed of an inner longitudinal layer of smooth muscle fibres and an outer circular layer, the two being separated by a layer of fibrous and elastic tissue.

2. The **male urethra** or **urogenital sinus** possesses a **stratified squamous epithelium** in its prostatic part, — a **stratified columnar** in its membranous portion, — a **simple columnar** in the spongy part, — a **stratified squamous** again in the fossa navicularis. The **tunica propria** possesses numerous elastic fibres and bears papillae particularly numerous in the fossa navicularis. Simple branched tubular glands, **Littré's glands,** occur in the whole urethra. The **muscular coat,** consisting of an inner longitudinal and an outer circular layer, occurs in the prostatic and the membranous divisions, but is absent in the spongy portion excepting a few, scattered, longitudinal fibres. **Blood-vessels** are numerous in the mucosa, and **lymphatics** in the submucosa. **Nerves** accompany the blood-vessels.

THE REPRODUCTIVE SYSTEM.

A. THE MALE REPRODUCTIVE ORGANS.

I. The **testicles** are compound tubular glands. Each is enclosed in a fibrous capsule, the **tunica albuginea**, which is thickened on the posterior side, forming the **mediastinum** or **corpus Highmori**. Fibrous **septa** extend between this and the tunica albuginea, thus dividing the testicle into **pyramidal lobules**, which contain the **seminiferous tubules**. The capsule exhibits two regions, an inner of loose fibro-elastic tissue rich in blood-vessels, the **tunica vasculosa**, and an outer, denser portion covered on its free surface with endothelium from the visceral layer of the **tunica vaginalis.**

1. The **seminiferous tubules** exhibit three regions, the convoluted tubules, which unite with one another and connect with the straight tubules, and these with the rete testis, which lies in the mediastinum.

a. The **convoluted tubules** are round canals lined with several layers of **epithelial cells** resting on a thin **basement membrane** and surrounded by several layers of **endothelioid connective-tissue cells.** The loose connective tissue between the tubules contains groups of spherical or polyhedral cells, the **interstitial cells.** When at rest the epithelial lining consists of nearly spherical cells, — when active these cells exhibit different stages in the process of **spermatogenesis.** The layer resting on the basement membrane contains two kinds of cells, the **sustentacular cells** or **Sertoli's columns** and the **spermatogenetic cells.** The latter by indirect division produce the next layer, which consists of larger cells, the **mother cells**; these divide twice, giving rise to four cells, the **daughter cells**, known as **spermatids** or **spermatoblasts,** arranged in a radial row. From them the **spermatozoa** are formed and during the development they become grouped about the end of Sertoli's columns, which have elongated radially. Whether the whole spermatozoon develops from the nucleus, or whether the head develops from the nucleus and the tail from a part of the protoplasm is still in dispute.

b. The **straight tubules** or **tubuli recti**, about 25 μ in diameter, are lined with a single layer of low columnar cells.

c. The **rete testis** consists of tubules of various sizes, lined with flattened epithelium.

The larger **blood-vessels** lie in the mediastinum and the tunica vasculosa, send branches to the septa and terminate in capillary networks around the tubules. The lymphatics and nerves have a similar distribution, but the method of nerve-termination is doubtful.

2. The **spermatozoa** consist of a head and a tail. The head, from 3 to 5 μ long and 2 to 3 μ broad, is flattened; the broad surface is oval, the narrow surface pyriform, attached at its broader end. The tail consists of three regions: — the middle piece, 4 to 6 μ long, to which the head is attached; the main piece, 40 to 50 μ long; and the end piece, about 10 μ long, consisting of the axial fibre only, which forms the axis of the entire tail.

II. The **excretory ducts** consist of the tubuli recti and the rete testis, which are within the testicle, of the vasa efferentia and the vas epididymis, which are within the epididymis, and of the vas deferens, its terminal expansion, the ampulla, the seminal vesicle and the ejaculatory duct.

1. The **vasa efferentia**, from 10 to 15 in number, are lined with a layer of ciliated columnar epithelium with a second layer of small cells between them and the basement membrane. In the initial portion of the tubules, the columnar cells are arranged in groups, separated by groups of cuboidal cells; the latter are not ciliated. The epithelium rests on a basement membrane surrounded by several layers of circularly-arranged, smooth muscle-fibres, and outside this, loose connective tissue. In the **globus major** of the epididymis, the convolutions of the vasa efferentia form conical lobules, the **coni vasculosi**, and unite to form the vas epididymis.

2. The **vas epididymis** consists of a stratified ciliated columnar epithelium, — generally composed of two layers of cells, — a basement membrane, circular smooth muscle-fibres, and loose connective tissue.

In the epididymis lie several tubules, — generally closed, — which are the remains of foetal organs. These are the **paradidymis** or **organ of Giraldès**, the **vas aberrans Halleri**, usually opening into the terminal portion of the vas epididymis, the **sessile hydatid** or **hydatid of Morgagni**, and the **stalked hydatid**. They are lined in general with cuboidal or columnar ciliated epithelium and contain a clear fluid.

3. The **vas deferens** consists of stratified non-ciliated columnar epithelium, a basement membrane, a tunica propria, a submucosa containing a few longitudinal smooth muscle-fibres in its initial portion, and two relatively thick layers of smooth muscle-fibres, an inner circular and an outer longitudinal.

4. The **ampulla** of the vas deferens has a similar structure, but the layers are thinner, and small, branched tubular glands are present in the mucosa.

5. The seminal vesicle has the same structure as the ampulla.

6. The ejaculatory duct consists of a simple columnar epithelium, a tunica propria, a thin submucosa, and a thin muscular coat consisting of an inner circular and an outer longitudinal layer.

III. The accessory glands are the prostate and Cowper's glands.

1. The prostate is enclosed in a fibrous capsule and consists of a number of branched tubular glands, whose tubules are surrounded by a small amount of fibrous and elastic tissue and by a large amount of smooth muscular tissue. The tubules are lined with a layer of columnar epithelium, separated from the basement membrane by a layer of small, rounded cells; they unite to form from 12 to 20 excretory ducts all opening into the urethra in the grooves on either side of the colliculus seminalis. In elderly men small, rounded concretions showing concentric lamellae occur in the tubules; these are the prostatic crystals.

The arteries of the prostate are branches of the vesical, haemorrhoidal and pudic arteries. They form capillary networks about the tubules, and the veins form a plexus in the capsule and connect with the dorsal vein of the penis and with branches of the internal iliac. Lymphatics follow the veins. The nerves, both medullated and non-medullated, are derived from the hypogastric plexus; their termination is unknown except in the superficial nerves, where Pacinian bodies occur.

2. Cowper's glands are compound tubular glands. The tubules, lined with a single layer of clear columnar cells, unite to form the excretory ducts, — one to each gland, — which are lined with two or three layers of cubical epithelium.

IV. The semen as ejected, consists of the products of the testicles, — composed of spermatozoa in a concentrated fluid, — diluted with secretions of the seminal vesicles, the prostate gland, and Cowper's glands. The prostatic fluid is opalescent with an acid reaction. The secretion of Cowper's glands is a clear, viscid fluid.

V. The penis consists of the corpora cavernosa, the corpus spongiosum enclosing the urethra and terminating in the glans penis, and the enveloping fascia and integument.

1. The corpora cavernosa consist of erectile tissue enclosed in a fibrous sheath, the tunica albuginea, about 1 mm thick and consisting of an outer layer of longitudinal bundles of white and elastic fibres surrounding both corpora, and an inner layer of circular fibres enclosing each corpus and thus forming along the median line the pectinate septum. The erectile tissue consists of anastomosing trabeculae of fibrous tissue connecting with the tunica albuginea, and containing bundles of smooth muscle-fibres. The surfaces of the trabeculae are covered with endothelium, which thus lines the intertrabecular cavernous or venous spaces. Into these venous spaces some of the smaller arteries open directly, the others open by means of capillaries. The arteries which supply the trabeculae without opening into the venous spaces are known as the helicine arteries. The chief arteries of the corpora cavernosa are branches of the pudic arteries and of the dorsal artery of the penis. From the cavernous spaces the blood passes to the pudendal veins and the dorsal vein.

2. The corpus spongiosum is essentially the same in structure as the corpora cavernosa, but the tunica albuginea is thinner, consisting of circular bundles of fibrous tissue and much elastic tissue. The trabeculae are smaller, and the arteries do not open directly into the venous spaces. A submucosa, rich in blood-vessels, connects the erectile tissue and the mucosa of the urethra. The glans consists essentially of a fine plexus of veins.

3. The integument of the penis is thin, dark in color, and attached by a very loose subcutaneous tissue. Fat is absent, and hairs occur near the attached end only. At the edge of the prepuce it takes on the structure of a mucous membrane. On the glans it is firmly united to the underlying fibrous tissue, which is beset with numerous papillae. At the base of the glans numerous modified sebaceous glands occur, Tyson's glands, whose secretion contributes the greater part of the smegma.

The lymphatics occur as a superficial set, — chiefly in the subcutaneous tissue and on the glans and the prepuce, — which pass to the inguinal glands, and also as a deep set, passing from the erectile tissue to the lymphatics in the pelvis. The nerves arise from the branches of the pudic nerve, terminating in the integument and in the mucous membrane, and from the sympathetic hypogastrio plexus, terminating in the erectile bodies. Genital corpuscles and Pacinian bodies are numerous in the glans.

B. THE FEMALE REPRODUCTIVE ORGANS.

I. The ovary is covered with a simple, low, columnar epithelium, the germinal epithelium, which contains occasional, larger, spherical cells, the primordial ova. The transition from the columnar epithelium to the endothelium of the adjacent serous membrane is marked by a sharp line. The epithelium encloses the connective tissue stroma, which is condensed just beneath the epithelium into a more compact layer, the tunica albuginea; this does not, however, represent a distinct membrane.

1. The stroma exhibits two regions, — an outer, the cortex, which contains the glandular tissue, occupying about a third of the diameter

of the ovary, and the medulla or central portion, which contains the larger blood-vessels and bundles of smooth muscle-fibres. The cells of the stroma are very numerous and spindle-shaped; amongst them are occasional groups of rounded cells, which resemble the interstitial cells of the testicle. Fibrous tissue is most abundant in the medulla.

2. The glandular tissue consists of ova surrounded by epithelial cells, which have grown into the stroma from the germinal epithelium. Only during foetal life and for a short time after birth are ova formed from the germinal epithelium. At first the egg-cells lie in rows surrounded by epithelial cells, constituting the primary egg tubes. These tubes break up into groups of cells, forming the nests of ova. Later each ovum, surrounded by a layer of epithelial cells, becomes separated from the others, constituting a Graafian follicle.

3. The Graafian follicles, of which the human ovaries contain some 70,000, consist at first of a single layer of flattened cells, the follicular epithelium, enclosing the ovum. The epithelial cells increase in size and number, become first cuboidal, then columnar, and finally form several layers of polyhedral cells; the ovum lies eccentrically. In the midst of these cells a cavity appears, filled with a fluid, the liquor folliculi, derived partly from the surrounding blood-vessels and partly from the cells themselves. The follicle constantly increases in size until it occupies the whole thickness of the cortex, being from 0.5 to 12 mm in diameter. It is enclosed in a connective-tissue sheath, the theca folliculi, consisting of an inner layer, the tunica propria, containing numerous cells and blood-vessels, and an outer layer, the tunica fibrosa, a fibrous layer. The follicular epithelium is known as the membrana granulosa, and that part of it which surrounds the ovum, as the discus proligerus or cumulus ovigerus. The cells adjoining the ovum are arranged radially and constitute the corona radiata.

4. The ovum increases in size, and the protoplasm becomes modified into a granular deutoplasm, or nutritive material, except a thin layer at the periphery and around the nucleus. The nucleus and the protoplasm constitute the vitellus.

The eccentrically placed nucleus is called the germinal vesicle, and contains the germinal spot, a deeply stainable body. Around the ovum a radially-striated membrane develops, the zona pellucida, separated from the vitellus by a narrow space, the perivitelline space. The mature ovum is about 0.2 mm in diameter. When the Graafian follicle is mature, it ruptures at the surface of the ovary, and the ovum, enclosed in the discus proligerus, passes into the abdominal cavity.

5. The corpus luteum is the ruptured Graafian follicle, and if pregnancy does not follow the escape of the ovum, it soon disappears, and is called the **false corpus luteum**; if pregnancy occurs, it persists for a long time and is known as the **true corpus luteum.** The cavity of the follicle becomes filled with blood from the ruptured vessels of the tunica propria, and the walls become folded and increase in thickness by the hypertrophy of the cells of the follicular epithelium, and by an ingrowth of blood-vessels and cells from the tunica propria. Ultimately the cavity of the corpus and the foldings of its walls become almost obliterated, and it appears as a mass of yellowish cells enclosed in a fibrous framework. Many Graafian follicles never reach maturity, and many mature follicles degenerate by disintegration and the action of leucocytes.

The **arteries**, branches of the ovarian and the uterine arteries, enter at the attached border of the ovary, pass through the medulla to the cortex, and form capillary networks about the Graafian follicles. The **veins** have a similar course.

Lymphatics are numerous in the medulla and occur in the theca of the follicles. The **nerves**, medullated and non-medullated, accompany the blood-vessels and form a dense plexus of gray fibres about the follicles.

6. The **parovarium** or **epoöphoron** is a series of tubules lying transversely in the broad ligament between the ovary and the Fallopian tube, and opening into a longitudinal tubule at the ends near the Fallopian tube. They are lined with low columnar ciliated epithelium. The **paroöphoron** consists of branched tubules, having a similar structure, which lie in the broad ligament between the ovary and the uterus.

Near the fimbria of the Fallopian tube, the **stalked hydatid of Morgagni** frequently occurs, and has a lining similar to that of the other tubules. All these structures are the remains of foetal organs.

II. The **Fallopian tube** or **oviduct** consists of a mucous, a muscular and a serous coat.

1. The **mucosa**, which possesses numerous **longitudinal folds**, consists of **simple ciliated columnar epithelium** resting on a fibroelastic tunica propria, which contains numerous cells and a thin **muscularis mucosae** of longitudinal, smooth muscle-fibres. There is a thin submucosa.

2. The **muscular coat** consists of a thick, inner, circular layer and a thin, outer, longitudinal layer of smooth muscle-fibres.

3. The **serous coat** consists of loose fibrous tissue covered with endothelium.

Blood-vessels are numerous in the mucosa, and are accompanied by lymphatics and nerves. The method of termination of the nerves is unknown.

III. The uterus exhibits a structure which varies according to the physiological condition of the organ, and represented by the resting uterus, the menstruating uterus, and the gravid uterus.

1. The resting uterus consists of a mucosa, a muscular and a serous coat.

a. The mucosa presents a smooth surface except on the lower third of the cervix, which is beset with small papillae. It consists of a simple ciliated columnar epithelium except on the papillae, which are covered with stratified squamous epithelium; the tunica propria consists of some fibrous tissue and numerous connective-tissue cells. It contains numerous, convoluted, simple, tubular glands lined in all cases with simple ciliated columnar epithelium; they extend as far as the muscular coat. In the cervix, in addition to the tubular glands, mucous crypts occur. They are macroscopic but small and by the retention of their secretion become distended cysts, the ovula Nabothi. The inner muscular coat is a very thick layer of smooth muscle-fibres representing the hypertrophied muscularis-mucosae. The fibres are arranged circularly in the lower part of the uterus and the cervix, but at the cervix there is an additional inner layer, which is longitudinal; in the upper part they are arranged concentrically about the openings of the Fallopian tubes. A submucosa is absent.

b. The muscular coat proper consists of two thin layers of irregularly disposed fibres; in the inner layer they are, in general, circularly arranged and in the outer, longitudinally.

c. The serous coat has the same structure as elsewhere, *i. e.* fibro-elastic tissue covered with endothelium.

The larger arteries lie in the outer part of the muscularis mucosae and send branches through the tunica propria, forming a capillary network beneath the epithelium and around the glands. The veins have a similar course and are very large. Lymphatics form a network in the mucosa and connect with networks in the muscular layers. The nerves are numerous, both medullated and non-medullated, and form networks throughout the uterus.

2. In the menstruating uterus there occurs first a thickening of the mucosa due to cell proliferation and the distention of the blood-vessels, followed by the disintegration of the epithelial lining, the rupture of the blood-vessels and the disintegration of the whole mucosa as far as the muscular layer. From the cells between the fibres of the inner part of the muscularis mucosae, the mucosa is

regenerated, and from the ends of the glands lying in the same region, the new glands and the whole uterine epithelium are formed. These changes do not involve the cervix, but pass progressively from the cervical end of the uterus towards the fundus.

3. The gravid uterus exhibits a great increase in the thickness of the muscular layers, produced by an increase in the length and breadth of the individual fibres, and in the number of fibres as well. The mucosa also thickens, and we recognize three regions, the decidua serotina, or part to which the ovum is attached, the decidua vera, or the rest of the uterine lining, and the decidua reflexa, or the part which grows around the ovum and disappears at about the middle of pregnancy. The thickened mucosa exhibits two portions, — a superficial compact layer, and a deeper, thicker cavernous layer, whose cavities are produced by the distended tubules of the glands. The epithelium of the uterine cavity and of the glands early disintegrates, and the orifices of the glands become closed. The arteries become distended and, after a spiral course in the mucosa, open at the surface of the decidua, thus bathing the villi of the chorion, — the outer portion of the foetal placenta, — with the maternal blood. The veins open in the same manner but do not have a tortuous course in the mucosa. In the mucosa occur numerous connective-tissue cells, multinucleated giant-cells, and decidual cells, which are of various shapes, sometimes uninuclear, sometimes multinuclear, and of a brown colour.

IV. The vagina consists of a mucous, a muscular, and a fibrous coat.

1. The mucous membrane consists of stratified squamous epithelium and a tunica propria of fibrous and elastic tissue, containing numerous leucocytes and beset with papillae. In the rugae smooth muscle-fibres occur. There is a submucosa of loose fibro-elastic tissue. Glands are absent. The hymen is a fold of the mucous membrane.

2. The muscular coat consists of smooth muscle-fibres arranged, in general, as an inner circular and an outer longitudinal layer, connected by numerous oblique fibres.

3. The fibrous coat consists of dense fibrous and elastic tissue, compact on the anterior surface, loose on the posterior.

The blood-vessels and lymphatics lie in the submucosa and send branches to the mucosa and the muscular layers. The veins are so large and numerous in the submucosa near the orifice of the vagina, that it has the structure of erectile tissue. The nerves form a plexus in the fibrous coat, and genital corpuscles occur in the mucosa.

V. The genitalia.

1. The vestibule and the labia minora or nymphae are covered

with stratified squamous epithelium resting on a fibro-elastic
tunica propria, which contains numerous mucous glands in the
region of the vestibule, and sebaceous glands on the nymphae.
In many places the submucosa consists of erectile tissue. The
glands of Bartolin, homologous with Cowper's glands, are
branched tubular glands, each supplied with a duct, opening at
the inner, posterior end of the nymphae. The ducts are lined
with simple cuboidal epithelium. Between the nymphae and the
labia majora lies a mass of erectile tissue on each side, about 2.5
cm. long, the bulbi vestibuli.

2. The clitoris has essentially the same structure as the penis
but only slightly developed; it consists of two corpora cavernosa
and a rudimentary glans and prepuce. In the glans, genital corpuscles occur, and sebaceous glands are found at its base and
on the prepuce.

3. The labia majora are folds of the integument. The outer surface has the same structure as the integument, the inner resembles
mucous membrane; adipose tissue is abundant.

VI. The mammary glands consist each of from fifteen to twenty
branched tubular glands, opening by as many ducts at the end of
the nipple; these ducts are enlarged just before reaching the nipple
to form the ampullae, which constitute temporary reservoirs for the
milk. Each of the fifteen to twenty glands constitutes a lobe, which
is divided into lobules.

1. The ducts and tubules are lined with a single layer of epithelium, which in the closed ends of the tubules and in the
smaller ducts consists of flattened or cuboidal cells, in the larger
ducts, of cuboidal cells, and in the excretory ducts and ampullae,
of columnar cells; near the external orifice of the ducts the
epithelium becomes stratified squamous. The epithelium rests on
a basement membrane surrounded by a circular layer of fibrous
tissue. The lobes and lobules are separated from one another by
areolar tissue and much adipose tissue.

2. In the nipple smooth muscle-fibres occur, arranged radially from
base to apex and circularly. A large amount of pigment is present
in the integument of the nipple and the areola. At the base of
the nipple and in the areola are small elevations, which mark the
orifices of accessory glands, the glands of Montgomery. In the
areola, sweat glands and sebaceous glands are also numerous.

The blood-vessels and lymphatics form capillary networks around
the tubules and the ducts. Nerves surround the tubules but are
more numerous in the superficial parts of the gland, and, in the
nipple, tactile corpuscles and Pacinian bodies occur.

In men and children the mammary glands consist mainly of connective tissue, excretory ducts, and solid rods of cells representing the tubules. In elderly women the tubules disappear.

Milk consists of a clear fluid, the **milk-plasma**, in which numerous oil-globules are suspended; as these do not coalesce, a thin membrane of caseine is supposed to surround them. Milk secreted before and for a few days after parturition contains in addition nucleated cells containing fat globules; these cells are known as **colostrum corpuscles.**

THE CENTRAL NERVOUS SYSTEM.

I. The membranes or meninges of the spinal cord and the brain are the dura, the arachnoid, and the pia. The arachnoid and the pia are sometimes considered as one membrane and called the pia.

1. The dura or outermost membrane consists on the cord of a dense layer of fibrous and elastic tissue, the fibres lying longitudinally, and contains numerous flattened connective-tissue cells and plasma cells. The inner surface is lined with endothelium. On the brain the dura consists of two layers, an inner, homologous with the dura of the cord, and an outer, which forms the periosteum of the adjacent surface of the cranium. The two layers have the same structure as the dura of the cord, but the fibres of the outer layer lie transversely to those of the inner layer. The dura is separated from the arachnoid by the subdural space, across which a few bundles of fibres pass to connect the two membranes. In some places epidural spaces, *i. e.* between the dura and the cranium, are present, with an endothelial lining. Blood-vessels and nerves are not numerous in the dura.

2. The arachnoid is a delicate membrane of loose fibrous tissue connected with the pia by numerous trabeculae, which traverse the relatively broad subarachnoid space. The surfaces and the trabeculae of the arachnoid are covered with endothelium. Villi occur on the outer surface, and are particularly numerous near the superior longitudinal sinus. They often become hypertrophied, forming the pacchionian bodies, which produce an elevation in the dura and often a corresponding depression in the cranium. They frequently contain calcareous concretions, the "brain sand." Blood-vessels and nerves are absent in the arachnoid.

3. The pia consists of two layers, an outer, highly vascular, more developed on the cord than on the brain, and an inner, closely attached to the nervous tissue, into which it sends connective-tissue fibres. The outer surface is covered with endothelium. The pia is composed of fibro-elastic tissue containing blood-vessels, lymph-clefts, and a few nerve-fibres; pigment cells occur in the outer layer. It is the only one of the three membranes which enters the fissures and the sulci.

The telae choroideae and the plexus choroideae are vascular villi on the folds of the pia, which project into the ventricles. They are covered with flattened or cuboidal epithelium.

II. The spinal cord consists of a gray substance composed mainly of nerve-cells, a white substance, composed mainly of nerve-fibres, and a supporting substance. The cord is partially divided by the anterior median fissure and by the posterior septum. In a cross section of the cord the gray substance has the shape of a letter H, the ends of which constitute the anterior and the posterior cornua; the connecting piece, the gray commissure, contains the central canal, so that an anterior and a posterior gray commissure are distinguished. The white substance is thus divided by the gray substance into three regions, the anterior, the lateral, and the posterior columns. The conus medullaris, the end of the cord, is almost entirely gray substance.

1. The gray substance varies in amount more or less as the size of the cord varies; it is greatest in the lumbar enlargement.

The cells in the anterior cornua are arranged in groups, whose position is indicated by their names, *i. e.* the median anterior, the lateral anterior, and, in the cervical and the lumbar enlargements, the lateral posterior in addition. The anterior roots of the spinal nerves emerge from the anterior surface of the anterior cornua.

The posterior cornua exhibit several regions; on the median side and adjacent to the posterior commissure lies the column of Clarke; in the same region laterally is a gray network, the reticular process. Posterior to this region are three masses known successively as the substantia gelatinosa Rolandi, the zona spongiosa, and the zona terminalis. The posterior roots of the spinal nerves enter on the median side of the posterior cornua.

Of the cells of the first type the largest are the motor cells, which lie in the anterior cornua. They are from 65 to 135 μ in diameter and have numerous long dendrites. The axis-cylinder process receives a medullary sheath just before leaving the cord, and emerges as a fibre of the anterior root of a spinal nerve. All of the fibres of the anterior root arise in this way; the motor cell is on the same side of the cord as the fibre.

The remaining cells of the first type, constituting the greater part of the gray substance, are of varying sizes, generally not more than half as large as the motor cells, and with long dendrites, which are less numerous and less branched than those of the motor cells. The axis-cylinder process may send off numerous collateral fibres in the gray substance and then entering the anterior or the lateral white column divide into an ascending and a descending branch; these branches run parallel to the long axis of the cord and send off lateral branches, which enter the anterior cornu and, dividing greatly, enclose the motor cells. These cells have been called column-cells. The axon of similar cells, instead of entering the white columns on the same side of the cord, may pass through the gray commissure to the opposite side of the cord; these are the commissural cells.

The cells in the column of Clarke are generally larger than those just described, and the axis-cylinder process passes to the lateral white column, where it does not divide, but goes to the cerebellum. The axis-cylinders of all nerves of the first type are surrounded by a medullary sheath.

Nerve cells of the second type occur in the posterior cornua and are known as interior or Golgi cells; their processes do not leave the gray substance.

The substantia spongiosa consists mainly of nerve-fibres crossing one another in various directions. The substantia gelatinosa consists almost entirely of supporting substance. The gray commissures consist of medullated nerve-fibres, and owe their colour to the supporting tissue. The central canal is surrounded by the substantia gelatinosa centralis and is lined with simple columnar epithelium, which is ciliated in children. In the adult the lumen of the canal may be obliterated.

2. The white substance consists of medullated fibres without the sheath of Schwann. The fibres are of various sizes, the largest being found in the anterior columns, the smallest in the median part of the posterior. Most of the fibres are parallel to the long axis of the cord, — some are oblique as in the white commissure, which lies anterior to the gray commissure and connects the two anterior white columns. The anterior and the lateral columns consist of fibres from the column-cells and from the cells of the brain; the posterior columns consist of fibres from the posterior roots of the spinal nerves. These fibres are arranged in more or less definite longitudinal tracts.

a. The anterior column has two tracts:

The direct pyramidal tract, or column of Türck, a descending tract, adjoins the anterior median fissure.

The anterior ground bundle lies between the direct pyramidal tract and the anterior cornu.

b. The lateral column exhibits some five tracts:

The direct cerebellar, an ascending tract, adjoins the periphery of the cord and lies anterior to the posterior cornu, extending about half way between the posterior and the anterior cornua.

The ascending antero-lateral tract, or tract of Gowers, occupies the periphery between the anterior end of the direct cerebellar tract and the anterior cornu.

The crossed pyramidal tract, a descending tract, lies between the direct cerebellar tract and the posterior cornu.

The descending antero-lateral tract lies anterior to the crossed pyramidal tract and adjoins the ascending antero-lateral tract. The mixed lateral tract adjoins the outer border of the gray substance.

c. The posterior column has two tracts separated from one another by a fibrous septum. Both are ascending tracts. The tract of Goll is triangular in cross section and adjoins the posterior median septum, extending from the periphery to the gray commissure.

The tract of Burdach lies between the tract of Goll and the posterior cornu.

The nerve-fibres which enter the cord from the cells of the ganglia of the posterior roots of the spinal nerves pass in part to the zona terminalis and in part to the posterior columns. They divide into two branches, an ascending and a descending, which give off lateral branches that enter the gray substance and terminate in numerous fibrils, in part in the substantia gelatinosa Rolandi, in part in Clarke's column, and in part in the anterior cornu where they enclose the motor cells.

3. The supporting substance is of two kinds, the fibrous connective tissue and the neuroglia.

a. The fibrous tissue enters from the pia and penetrates the white substance only, enclosing the blood-vessels.

b. The neuroglia or nerve-cement consists almost exclusively of cells, the glia-cells, which have the same embryonic origin as the nerve-cells. It forms a subpial layer, accompanies the fibrous tissue into the cord, and closely encloses the nerve-fibres, and occurs to a less extent in the gray substance. The glia-cells are of two kinds, the cells of the ependyma, and Deiters' cells.

The ependymal cells constitute the epithelial lining of the central canal, and their processes extend for some distance into the substance of the cord.

Deiters' cells, of various shapes, possess very numerous branched processes. There are two types with intermediate forms, the mossy-cells with greatly branched processes, occurring chiefly in the gray substance, and the spider-cells, many of whose processes are longer and not greatly branched, occurring in the white substance. In the substantia spongiosa centralis Deiters' cells possess long, stiff processes, in the zona spongiosa of the posterior cornu, the processes are very numerous and delicate.

The arteries of the cord enter at the periphery, and through the anterior median fissure from the arteria sulci. Those from the periphery supply for the most part the white substance, — those from the fissure, the gray substance. In general, the **veins** have a similar course.

III. The **brain** consists of the same general structures as the cord, *i. e.* gray, white, and supporting substance.

1. The **medulla oblongata** differs from the spinal cord chiefly in the arrangement of its gray substance.

a. The **posterior gray substance** exhibits three regions on each side of the median line, — the continuation of the posterior cornu of the cord, the **substantia gelatinosa Rolandi** lying laterally in the funiculus of Rolando, instead of posteriorly, and two new masses derived from the thickened base of the posterior cornu, — the **nucleus cuneatus** lying in the column of Burdach (the funiculus cuneatus), and the **nucleus gracilis,** lying in the column of Goll (the funiculus gracilis). As the central canal opens into the fourth ventricle, the gray substance which lay posterior to it becomes lateral; the **anterior cornua** are divided into three regions, — a median representing the bases of the cornua, lying on the floor of the ventricle in the **funiculus teres,** and two lateral, one on each side of the medulla, each constituting a **nucleus lateralis.** The intermediate area between these three masses is occupied by the **formatio reticularis,** which consists of nerve-cells and crossing nerve-fibres. Laterally in the funiculus teres on each side lies a mass of nerve-cells forming the **hypoglossal nucleus,** from which the roots of the hypoglossal nerve arise. Lateral to the hypoglossal nucleus is the common nucleus of the **pneumogastric,** the **glosso-pharyngeal,** and the **spinal accessory nerves.** In the olivary bodies is a wavy line of gray substance, the **corpus dentatum,** consisting of neuroglia and nerve-cells.

b. The four principal **tracts of white substance** are for the most part continuations of the tracts of the cord.

The **anterior pyramid** consists of fibres from the direct pyramidal tract and the crossed pyramidal tract; the decussation of the fibres of the latter produces the division of the anterior cornua into three masses.

The **lateral tract** consists of fibres from the entire lateral column including the anterior ground bundle, except the fibres of the crossed pyramidal and the direct cerebellar tracts.

The **posterior pyramid** lies at the lower end of the fourth ventricle and consists of a continuation of the column of Goll merging into the restiform body.

The **restiform body** consists of fibres from the direct cerebellar tract and from the posterior columns, together with fibres arising within the medulla itself.

2. The pons exhibits two regions, an anterior and a posterior.

   a. The anterior region consists chiefly of transverse nerve-fibres, the commissural tracts, connecting the hemispheres of the cerebellum, through which the bundles of the anterior pyramids of the medulla pass. The commissural fibres are arranged in three sets, — the anterior or superficial bundles, the middle bundles, and the posterior or deep bundles. Numerous ganglion-cells occur amongst these fibres.

   b. The posterior region consists mainly of continuations of the formatio reticularis and the gray substance of the floor of the fourth ventricle of the medulla. This gray substance forms the nuclei connected with the fifth, sixth, seventh, and eighth nerves. Lateral masses of pigmented nerve-cells, the **substantia ferruginea,** occur beneath the floor of the ventricle in the pons, and appear bluish-white when seen from the surface, forming the locus caeruleus.

3. The cerebellum consists of a white medullary substance enclosed by three layers of gray substance, — the granule layer, the layer of Purkinje's cells, and the molecular layer, the outermost.

   a. The granule layer, of a rust colour, consists of several layers of cells, which include a few glia-cells, a very large number of small granule-cells and some large granule-cells. The small cells possess short protoplasmic processes and an axis-cylinder process, which passes outward into the molecular layer where it divides into T-branches, which run parallel to the surface and terminate in free ends without branching. The large cells possess long and richly branching protoplasmic processes, which extend into the molecular layer, while the axis-cylinder process terminates in the granule layer by dividing many times ; these are therefore cells of the second type. Many medullated nerve-fibres pass through the granule layer from the white medullary substance, and form a layer at the outer boundary of the granule layer parallel to the surface, from which branches pass into the molecular layer.

   b. The cells of Purkinje form a single layer. They are very large, from 60 to 70 μ long. The axis-cylinder process passes through the granule layer into the white medullary substance ; it gives off collateral fibres, which sometimes return to the molecular layer. Two large protoplasmic processes are given off in the molecular layer, which branch greatly in a plane at right angles to the long axis of the convolution, — the longer branches extending for the most part to the surface of the molecular layer.

   c. The molecular layer contains two kinds of ganglion-cells.

The large cells lie in the deeper part of the layer; they send protoplasmic processes towards the surface, while the axis-cylinder process runs parallel to the surface and sends off collateral fibres, some of which branch very freely forming a basket-work around the body of the cells of Purkinje. The small ganglion-cells lie nearer the surface; they are multipolar. Medullated nerve-fibres occur in the molecular layer; they are chiefly continuations of fibres lying in the granule layer, and after losing the medullary sheath they terminate freely between the protoplasmic processes or near the nerve-cells, both in the outer and in the deeper parts of the layer.

d. Several nuclei, of nerve-cells, lie in the white medullary substance on either side of the median line. These are the dentate nucleus, forming an irregular wavy line, the nucleus emboliformis, a rounded mass connected with the dentate nucleus at its median extremity, and the nucleus globosus and nucleus fastigii (or nucleus of the roof), which are rounded masses lying still nearer the median line, and differing from the first two by possessing larger ganglion cells. Numerous nerve-fibres pass through these nuclei.

e. The white medullary substance consists of medullated nerve-fibres without a neurilemma, which form the arbor-vitae. They consist of the fibres of the superior and the inferior commissures of the vermis and of the superior, middle, and inferior peduncles of the cerebellum.

f. The supporting substance consists of fibrous connective tissue and neuroglia. In the outer portion of the granule layer are glia-cells with long processes, which extend through the molecular layer and terminate in conical enlargements at the surface; they also have short processes extending into the granule layer. Cells resembling mossy-cells occur throughout the gray substance, and spider-cells are found in the white substance.

4. The crura cerebri or cerebral peduncles exhibit an anterior portion, the crusta pedunculi, consisting of two separated crescentic masses, one on each side of the median line, composed of ascending and descending fibres, — and a posterior portion, the tegmentum, consisting of a prolongation of the posterior structures of the pons. The crusta is separated from the tegmentum by the substantia nigra, a mass of deeply pigmented ganglion-cells. On the floor of the aqueduct of Sylvius are groups of ganglion-cells, which form the nuclei connected with the third and the fourth nerves. In the formatio reticularis lie the red nuclei, one on each side of the median line and consisting of pigmented nerve-cells.

5. The corpora quadrigemina are known as the inferior or posterior pair and the superior or anterior pair.

    a. The inferior pair consist almost exclusively of gray matter forming a lenticular nucleus on each side of the median line, and are connected by a gray commissure. A thin layer of medullated nerve-fibres covers the outer surface of the gray substance.

    b. The superior pair exhibit four superposed layers : —

The stratum zonale, or outermost, is a superficial white layer of nerve-fibres running transversely, which are derived chiefly from the optic tract.

The stratum cinereum is a gray cap, crescentic in cross section, whose nerve-cells generally send their protoplasmic processes outwards, and their axis-cylinder processes inwards towards the deeper layers.

The stratum opticum consists of nerve-cells and nerve-fibres. The cells send.their axis-cylinder process into the next deeper layer. The fibres are derived largely from the optic tract and vary much in size.

The stratum lemnisci consists of both nerve-cells and nerve-fibres ; some of the latter are derived from the nerve-cells of this layer.

6. The geniculate bodies are closely connected with the corpora quadrigemina, — the external bodies with the superior, the internal bodies with the inferior.

    a. The external or lateral geniculate bodies consist of alternate bands of gray and of white substance, the latter consisting of fibres from the optic tracts.

    b. The internal or mesial geniculate bodies consist of nerve-cells and nerve-fibres ; the apparent connection of the latter with the fibres of the optic tract is probably not real.

7. The optic thalami consist chiefly of gray substance, through which several tracts of nerve-fibres pass. A vertical white septum, S-shaped in section, divides each thalamus into two parts, a long lateral nucleus, and a shorter median nucleus; the latter is further separated by a second white septum from the anterior nucleus, which lies in the anterior tubercle.

8. The subthalmic region consists of three layers known respectively as, —

    a. The stratum dorsale, consisting chiefly of longitudinal fibres and a continuation of the red nucleus of the crura cerebri, —

    b. The zona incerta, a reticulum, a prolongation of the tegmentum, —

    c. The corpus subthalmicum, or nucleus of Luys, a gray mass containing numerous nerve-cells and a plexus of fine nerve-fibres, and enclosed in a layer of white substance.

9. The pineal body or epiphysis in adult man consists of a number of tubules surrounded by connective tissue and lined with cuboidal epithelium. The tubules frequently contain concretions, the "brain sand," which occur also in other parts of the brain and its membranes. There are found also throughout the brain, bodies with concentric striations, the corpora amylacea.

10. The corpora mammillaria or albicantia consist of gray substance enclosed in white substance; the nerve-cells are arranged in each corpus in two masses, a lateral and a median; the nerve-fibres come chiefly from the fornix.

11. The tuber cinereum contains on each side adjoining the optic tract, a mass of gray substance, the basal optic ganglion, which probably has no connection with the optic nerve.

12. The pituitary body or hypophysis cerebri, which forms the extremity of the infundibulum, a hollow process in the middle of the tuber cinereum, consists of two lobes, a posterior and an anterior.

a. The posterior lobe consists of a connective tissue network and numerous stellate or spindle-shaped cells. Sometimes a cavity is present, lined with ciliated columnar cells.

b. The anterior lobe consists of a number of slightly convoluted tubules, lined with columnar epithelium and enclosed in connective tissue, which contains numerous blood-vessels and lymph-clefts. The tubules are often entirely filled with cells so that the cavity is obliterated. A colloid substance sometimes occurs in the lumen of the tubules. In structure the lobe resembles the thyroid body.

13. The cerebrum consists of a gray cortex, varying from 2 to 4 mm in thickness, which encloses the white medulla.

a. The cerebral cortex exhibits in general five layers merging more or less with one another.

The first, or the molecular layer, the outermost, is about one-tenth of the total thickness of the cortex, and consists chiefly of neuroglia, together with the protoplasmic processes of cells from deeper layers, and some medullated nerve-fibres running parallel to the surface, the tangential fibres.

The second, or layer of small pyramidal cells, has about the same thickness as the first. The cells have the form of pyramids with the apex turned towards the periphery. From the apex a long, branching protoplasmic process arises, which terminates in numerous branches in the molecular layer. From the sides of the base of the cells other protoplasmic processes come off. From the base arises the axis-cylinder process, which as a medullated fibre may pass to the molecular layer, thus forming a cell of the second type.

The third, or the **layer of large pyramidal cells**, is generally the thickest layer. The cells are like the cells of the second layer but larger, and the axis-cylinder process, after giving off collateral fibres, always goes to the white substance of the medulla. There are also cells of the second type.

The **fourth**, or the **layer of polymorphous cells**, is narrower than the third, and the cells are small and irregular in outline with numerous dendrites and an axis-cylinder process, which generally passes towards the white substance, but sometimes extends peripherally into the molecular layer, terminating there as a cell of the second type.

The **fifth**, or the **layer of spindle cells**, is thicker than the fourth and contains spindle-shaped and irregular cells. Bundles of **nerve-fibres** extend from the medulla through about half the thickness of the cortex as the **radial bundles**; they consist in part of medullated fibres from the pyramidal cells and in part of fibres from the white substance, which form the **tangential fibres** of the molecular layer and likewise the **superradial fibres**, transverse fibres lying in the second layer. The collateral fibres from the larger pyramidal cells constitute the **interradial fibres**.

The layers of the cortex vary more or less in different regions; the greatest variations occur in the **hippocampal region**, *i. e.* in the hippocampus major or cornu Ammonis and in the fascia dentata.

In the **hippocampus major** seven layers exist, which passing from the outside towards the ventricle are known as follows:

First, the **lamina medullaris involuta**, consisting largely of the white fibres of the molecular layer;

Second, the **stratum granulosum**, containing a number of small ganglion-cells, whose axis-cylinders are directed towards the periphery;

Third, the **stratum lacunosum** or **laciniosum**, consisting of axis-cylinder and protoplasmic processes;

Fourth, the **stratum radiatum**, consisting of the apical processes of the pyramidal cells of the next layer;

Fifth, the **stratum cellularum pyramidalium**, consisting of pyramidal cells, whose axis-cylinders pass into the deeper layers;

Sixth, the **stratum oriens**, corresponding in structure to the fifth cortical layer;

Seventh, the **alveus**, a thin layer of medullated nerve-fibres, parallel to the surface of the ventricle.

The **fascia dentata** exhibits four layers:

First, the **stratum marginale**, a thin layer of medullated nerve-fibres, continuous with the lamina medullaris involuta;

Second, the stratum moleculare, consisting chiefly of neuroglia and the processes of nerve-cells;

Third, the stratum granulosum, consisting of small pyramidal cells, whose axis-cylinders pass centrally;

Fourth, the nucleus fasciae dentatae, composed of nerve-fibres from the alveus and numerous ganglion-cells, consisting of large pyramidal cells, polymorphous cells, and spindle-shaped cells.

The septum lucidum, representing a rudimentary cortex, exhibits four layers:

First, a thin superficial layer of medullated nerve-fibres;

Second, a gray layer containing small pyramidal cells;

Third, a gray layer containing spindle-cells;

Fourth, a layer of medullated fibres corresponding to the white cerebral medulla.

b. The corpora striata consist of gray matter and nerve-fibres; the gray matter is present in each corpus as the nucleus caudatus and the nucleus lenticularis.

The nucleus caudatus is bounded above by the ependyma of the ventricle and below by the internal capsule. It contains multipolar cells and small cells of the second type, together with nerve-fibres coming from the internal capsule; these fibres form white striae in the gray substance.

The nucleus lenticularis is separated for the most part from the nucleus caudatus by the white internal capsule; the structure of the two nuclei is similar.

c. The olfactory lobes consist each of two regions of histological importance, the olfactory tract and the olfactory bulb.

The olfactory tract consists of a flattened central gray mass of neuroglia, surrounded by a layer of medullated nerve-fibres; this layer is thick beneath and encloses an oval area of gray substance above; the whole is surrounded by a thin layer of gray substance.

The olfactory bulb exhibits in cross section, passing from the superior to the inferior surface, the following structures: —

A very thin gray cortical layer;

A layer of medullated fibres forming a flattened ring, which encloses the mass of neuroglia;

A granule layer consisting of small nerve-cells, and large pyramidal cells, the mitral cells, whose axis-cylinders pass to the medullated fibres of the ring, and the protoplasmic processes end in free branches or in dense spheroidal masses of branches, which lie in the next layer;

A layer of olfactory glomeruli, which consist of these terminations of the mitral cells and the branched ends of olfactory nerve-fibres;

A layer of olfactory nerve-fibres, which are non-medullated.

d. The ventricles of the brain, including the aqueduct of Sylvius, are lined with neuroglia and ependymal cells, an extension of the lining of the central canal of the spinal cord. In young children the columnar ependymal cells are ciliated; the cilia disappear to a great extent in the adult.

e. The white substance or medulla of the cerebrum consists of medullated nerve-fibres without a neurilemma, forming, in general, three systems, the association fibres, the commissural fibres, and the projection fibres. Our knowledge of the arrangement of the nerve tracts in the medulla is still incomplete.

The association fibres connect convolutions in the same hemisphere, — either adjacent convolutions, or convolutions lying near one another, or convolutions widely separated. These longer tracts have been classified as follows:

The fasciculus uncinatus, connecting the inferior convolution of the frontal lobe with the uncinate gyrus of the temporal.

The fasciculus longitudinalis inferior, connecting the anterior part of the temporal lobe with the occipital.

The fasciculus longitudinalis superior, connecting the frontal lobe with the occipital and the superior part of the temporal.

The cingulum, lying above the corpus callosum in the cingulate convolution.

The fasciculus perpendicularis, connecting vertically the inferior parietal lobe with the fusiform.

The fornix, connecting the hippocampal region by means of the fimbria with the corpus mammillare, which is connected with the thalamus opticus by the fibres of Vicq d'Azyr.

The commissural fibres connect the two hemispheres and consist of the corpus callosum and the anterior commissure.

The projection fibres connect the hemispheres with other parts of the brain. They consist of the fibres of the crura cerebri, i. e. of the crusta and the tegmentum. The fibres of the crusta form in the hemispheres a part of the radiation in the medulla, known as the corona radiata; they are derived for the most part from the axis-cylinder processes of the cells of the cortex. The principal tracts which are continuous between the peduncles and the cortex, are the pyramidal tracts, the lateral tracts, and the fibres from the nuclei of the pons and the cerebellum.

f. The neuroglia of the cerebrum consists of ependymal cells and Deiters' cells. Spider-cells occur in the white substance, and mossy-cells in the gray substance often attached to the blood-vessels.

g. The blood-vessels form a fine network in the gray substance and a loose network in the white substance. They are surrounded by perivascular lmyph spaces connecting with the lymph spaces in the membranes. Numerous isolated lymph spaces occur around large ganglion-cells in the cerebral cortex and about some of the glia-cells.

THE SPECIAL SENSE ORGANS.

A. THE EYE AND ITS APPENDAGES.

I. The eyeball consists of three coats, — the tunica externa comprising the cornea and the sclera, — the tunica media comprising the iris, the ciliary body and the choroid, — the tunica interna or retina. Within are the aqueous humour, the lens, and the vitreous body.

1. The cornea consists of five layers as follows, — passing from the outside towards the inside:

a. The anterior epithelium, stratified squamous, consists of several layers of cells, — the outermost flattened, the deeper polyhedral, the deepest columnar.

b. The anterior limiting membrane, membrane of Bowman, or lamina elastica anterior, is apparently homogeneous; it is a modification of the surface of the next layer.

c. The substance proper, substantia propria, constituting the greater part of the cornea, consists of connective-tissue fibrillae united by cement substance into bundles and lamellae lying parallel with the surface. The plates are united by oblique fibres, the arcuate fibres. In the cement substance are irregular, branched spaces or lacunae, the corneal spaces, connected with one another by canaliculi. The spaces contain lymph and lymph corpuscles, and fixed, branched cells, the corneal corpuscles.

d. The posterior limiting membrane, membrane of Descemet, or lamina elastica posterior, is a clear, homogeneous membrane, thinner than the anterior limiting membrane.

e. The posterior endothelium consists of a single layer of flattened, polygonal cells.

Blood-vessels are absent in the cornea except at the margin. **Nerves** are very numerous, forming a ground plexus of non-medullated fibres in the midst of the substantia propria, a **subbasilar plexus** beneath the anterior limiting membrane, a **subepithelial plexus** and an **intraepithelial plexus.**

2. The sclera consists chiefly of fibrous connective tissue, arranged in interlacing bundles lying equatorially and meridionally. Elastic fibres occur and also branched, connective-tissue cells lying in lacunae. The sclera is covered on both the outer and the inner surface with endothelium. The innermost part of the sclera, known as the lamina fusca sclerae, consists of loose, fibro-elastic tissue and contains branched, pigmented cells; it is loosely attached to the lamina suprachoroidea, a similar layer on the outside of the choroid.

Blood-vessels are not numerous; the nerves form an interfascicular plexus.

The **episcleral space** or **space of Tenon** separates the sclera externally from the **capsule of Tenon**, a fibro-elastic membrane lined with endothelium and attached to the surrounding structures of the orbit. The capsule is essentially a synovial sac and communicates with the **perichoroidal space**, which lies between the sclera and the choroid.

3. The **choroid** consists of three layers, — the layer of large blood-vessels, the layer of capillaries, and the glassy membrane.

a. The **layer of large blood-vessels** or outer layer consists of a stroma composed chiefly of elastic tissue and pigmented cells, surrounding the large arteries and veins. Longitudinal smooth muscle-fibres accompany the largest arteries. The inner part of this **stroma layer** exhibits a narrow, non-pigmented zone, the **boundary zone**, in contact with the next layer.

b. The **choriocapillaris** or capillary layer, a thin layer, consists of close capillary networks imbedded in a homogeneous matrix.

c. The **glassy membrane** or vitreous lamina is a thin, homogeneous layer adjoining the pigmented layer of the retina.

4. The **ciliary body** consists of the ciliary ring, the ciliary processes, and the ciliary muscle.

a. The **ciliary ring** includes the portion of the choroid lying between the ora serrata and the ciliary processes. It differs from the choroid in that the stroma consists chiefly of fibrous instead of elastic tissue, and that muscle-fibres are present and the choriocapillaris absent.

b. The **ciliary processes** consist of a ring of from seventy to eighty radial elevations composed of fibrous tissue and blood-vessels and covered by a continuation of the glassy membrane. On the inner surface they are covered by the **pars ciliaris retinae** consisting of a layer of tall, columnar cells, — whose free surface is covered by a cuticula, the membrana limitans interna, — resting on a layer of low, pigmented cells, which adjoin the glassy membrane.

c. The **ciliary muscle** consists of three sets of smooth muscle-fibres:

Meridional fibres, nearly parallel with the sclera;

Radial fibres, adjoining the meridional;

Circular fibres, the ring-muscle of Müller, at right angles to the others.

5. The iris consists of five layers:

a. The anterior endothelium is a continuation of the posterior endothelium of the cornea and covers the anterior surface of the iris.

b. The anterior boundary layer consists of several layers formed of networks of connective-tissue cells and resembles adenoid tissue.

c. The vascular layer consists of a connective-tissue stroma, containing lymph-clefts and enclosing the blood-vessels and the smooth muscle-fibres. The radial arteries arise from the circulus iridis major and open near the margin of the pupil into the circulus iridis minor, from which capillaries connect with the veins. The muscles form two sets, a ring around the pupillary margin, the sphincter of the pupil, and radial bundles, the dilator of the pupil.

d. The posterior boundary layer or vitreous lamella is a clear, homogeneous, elastic membrane.

e. The pigment layer or pars iridica retinae consists of two layers, an anterior of spindle-shaped cells, and a posterior thicker layer; the cells of both layers are densely pigmented so that the boundaries are generally obscured. The free surface is covered by a cuticula, the membrana limitans iridis. In blue eyes no pigment exists in the stroma cells of the vascular layer; in darker eyes it is present there; in albinos no pigment is present in the iris.

The nerves form an irregular, non-medullated plexus in the anterior part of the iris.

6. The irido-corneal angle marks the junction of the cornea, the sclera, the iris, and the ciliary body. The cornea passes directly into the sclera, the transition being marked by an oblique line. At the periphery of the cornea, the posterior limiting membrane breaks up into a number of fibres connected with the iris and forming the ligamentum pectinatum iridis, which is only slightly developed in man. At the irido-corneal angle other fibres from the posterior limiting membrane, form trabeculae with fibres from the iris, from the attachment of the ciliary muscle and from the sclera, enclosing spaces, the spaces of Fontana, which are lined with endothelium and connect with the anterior chamber of the eye. In the sclera in this region is a flattened ring canal, the canal of Schlemm, a venous channel, having communication with the lymph-spaces of Fontana.

7. The retina exhibits three regions, the pars optica, the pars ciliaris, and the pars iridica. The pars optica lines the whole posterior chamber of the eye as far as the ora serrata and con-

sists of two laminae, an inner sensory, and an outer pigmented. The sensory part of the retina consists of an inner cerebral lamina, and an outer neuro-epithelial lamina adjoining the pigment layer.

The cerebral lamina exhibits five layers as follows:

a. The nerve-fibre layer, or innermost, consists of naked axis-cylinders. The layer is thickest at the entrance of the optic nerve and thinnest at its termination at the ora serrata. The fibres are mostly centripetal, arising from the ganglion-cells of the adjoining layer.

b. The ganglion-cell layer is a single layer of multipolar ganglion-cells, whose axis-cylinders pass into the nerve-fibre layer, and their protoplasmic processes into the next layer, — the inner reticular.

c. The inner reticular layer is composed of a fine network of supporting tissue and densely interlacing processes coming from all the ganglion-cells of the retina.

d. The inner granule or nuclear layer exhibits three kinds of ganglion-cells. The innermost layer is of multipolar cells whose branched processes terminate in the inner reticular layer, and from some an axis-cylinder process passes into the nerve-fibre layer. Then there are several layers of bipolar ganglion-cells, constituting the ganglion retinae, whose centripetal processes terminate in varicose branches in the inner reticular layer, while the peripheral processes pass into the next layer, — the inter-reticular, — and terminate in fine branches near the inner ends of the neuro-epithelial cells. Finally at the outer border of the inner granule layer lie stellate ganglion-cells known as spongioblasts or amacrines, whose protoplasmic processes pass into the adjoining layer; the course of the axis-cylinder processes is disputed; they are said to pass into the nerve-fibre layer, or to terminate in branches in the inner reticular layer, or, by others, to pass into the adjoining outer layer.

e. The outer reticular layer consists of supporting tissue, the processes of ganglion-cells and subepithelial ganglion-cells, whose processes are distributed precisely like those of the bipolar cells of the ganglion retinae.

The neuro-epithelial lamina consists of rod-visual cells and cone-visual cells. The lamina exhibits four layers:

a. The fibre-layer of Henle consists of the bases of the visual cells, forming a radially striated layer resting on the outer reticular layer.

b. The outer granule or nuclear layer contains the nuclei of the visual cells.

c. The membrana limitans externa is pierced by the visual cells and separates the outer granule layer from the layer of rods and cones.

d. The layer of rods and cones consists of the outer segments of the visual cells.

The rod-visual cells consist of two parts :

a. The inner segment or rod-fibre is a fine filament having an expanded base and an enlarged portion containing the nucleus, or rod-granule; the nucleus exhibits light and dark transverse bands.

b. The outer segment or rod, some 60μ long and 2μ in diameter, exhibits two regions : — (1) an enlarged inner portion, which is granular near the limiting membrane, and longitudinally fibrillated in its outer portion, or fibre-body; and (2) the homogeneous cylindrical rod, which is the exclusive seat of the visual purple or rhodopsin. There are supposed to be some 130 million rods in the human retina.

The cone-visual cells likewise consist of two parts :

a. The inner segment or cone-fibre is a broad filament with a pyramidal, expanded base resting on the outer reticular layer, and having a nucleated enlargement, the cone-granule, near the limiting membrane.

b. The outer segment or cone is shorter than the rod, but like it exhibits two regions : — (1) an enlarged inner portion having a fibre-body, and (2) the conical outer portion. Three and a half million cones are said to exist in the human retina.

The pigment layer consists of a single layer of cells, usually hexagonal in surface view, and exhibiting two regions, — an inner deeply pigmented, which adjoins the rods and cones, and an outer clear region containing the nucleus and adjoining the choroid.

The supporting substance of the retina consists chiefly of the radial fibres of Müller, which extend from the inner surface to the layer of rods and cones. The conically expanded inner ends of the fibres form the membrana limitans interna. The fibres then pass through the inner layers, and exhibit a nucleated enlargement in the inner granule layer, whence they pass to the membrana limitans externa, of which they form a part. In the outer nuclear layer they break up into a network of fibrils, while in all the layers which they traverse, they send off very numerous lateral processes. From the external surface of the membrana limitans externa, delicate processes, the fibre-crates, extend between the rods and cones. In the outer reticular layer, supporting cells resembling glia-cells occur, lying parallel with the surface of the layer; in some no nucleus is apparent.

The macula lutea exhibits a great increase in the number of cells in the ganglion-cell layer, so that they often form several layers, and an increase in the number of cells in the inner granule or nuclear layer. At the same time the rods disappear from the outermost layer. Towards the fovea centralis the layers of the cerebral lamina become thinner and fuse with one another until in the centre, the fundus foveae, nothing remains but the layer of cone-visual cells and one or two nerve-cells. Yellow pigment occurs in the cerebral layer but not in the neuro-epithelial layer.

At the ora serrata the nerve-fibre and the ganglion-cell layers cease, and also the layer of visual cells, — the rod-visual cells disappearing first. The two reticular layers terminate abruptly, and thus the two nuclear layers meet and become directly continuous with the columnar cells of the pars ciliaris retinae. Vacuoles often occur in old age in the retina near the ora serrata.

The larger blood-vessels of the retina are branches of the single artery which enters through the optic nerve, and lie in the nerve-fibre layer. From them two sets of capillaries arise, an inner network in the nerve-fibre layer, and an outer network in the inner nuclear layer. The arteries and the veins are surrounded by perivascular lymph-clefts.

8. The vitreous body is a clear fluid consisting of about 98.5% of water, and enclosed in the glassy hyaloid membrane. It contains a few very delicate fibres in its anterior part near the ora serrata, and cells, which are chiefly amoeboid. The hyaloid canal, bounded by a delicate membrane, passes from the optic papilla to the posterior part of the capsule of the lens.

9. The lens consists of two structures, the lens-epithelium and the lens-fibres, and is enclosed in the lens-capsule.

a. The lens-epithelium consists of a single layer of cuboidal cells. It covers the anterior surface as far as the equator, where it merges into the lens-fibres.

b. The lens-fibres consist of greatly elongated cells resembling hexagonal prisms and united by a small amount of cement substance. The fibres near the periphery of the lens are nucleated and exhibit smooth outlines ; in the centre they are not nucleated, and the outlines are finely serrated. The ends of the fibres meet at the anterior and the posterior poles of the lens along radiating lines, which constitute the anterior and posterior lens-stars.

c. The lens-capsule is a transparent elastic membrane, about twice as thick on the anterior as on the posterior surface of the lens.

d. The suspensory ligament of the lens, the zone of Zinn or

zonula ciliaris, consists of fibres from the hyaloid membrane of the vitreous body arising as far back as the ora serrata and closely attached to the surfaces of the ciliary processes. From the depressions of these processes, the fibres extend to the anterior part of the lens-capsule ; from the elevations they pass to the equatorial and the posterior parts of the capsule. Thus a saccular space is formed around the periphery of the lens, bounded anteriorly by the ligament, posteriorly by the vitreous body, and centrally by the equatorial surface of the lens. This space is the **canal of Petit.**

The **blood-vessels** of the eyeball consist of two distinct systems, the **retinal** and the **ciliary**; they are connected by a few branches at one point only, the point of entrance of the optic nerve.

The **lymphatics** consist of spaces only, constituting two systems, the anterior and the posterior.

1. The **anterior lymphatics** comprise the spaces of the cornea and the sclera, the anterior chamber (the aqueous humor), the posterior chamber, and the canal of Petit.

2. The **posterior lymphatics** comprise the hyaloid canal, the perivascular spaces of the retinal blood-vessels, the clefts of the optic nerve, the perichoroidal space, Tenon's space, and the supravaginal space enclosing the optic nerve.

The **ciliary nerves**, which supply all of the eyeball except the retina, enter near the optic nerve, pass between the sclera and the choroid giving off branches, and form the **ciliary ganglionic plexus** on the ciliary body ; from this plexus, branches pass to the anterior parts of the eye.

II. The **optic nerve** is invested with the same membranes as the central nervous system, — a dura, an arachnoid, and a pia, with subdural and subarachnoidal lymph-spaces. The fibres composing the nerve-fibre bundles are medullated and without a neurilemma. Mossy glia-cells enclose the bundles. Where the optic nerve enters the eyeball, the dura becomes continuous with the outer portion of the sclera, — the arachnoid terminates in fibres, — the pia fuses with the inner portion of the sclera, which, with a part of the choroid, is here pierced by openings through which the fibres of the optic nerve pass, and is known as the **lamina cribrosa.** At this point the medullary sheath ceases. In the axis of the distal part of the optic nerve lie the retinal artery and vein.

III. The **eyelids** exhibit five layers in sagittal section : —

1. The **integument** covers the outer surface and the free border. It is thin and possesses fine hairs, small sweat glands, and, in the corium, pigment cells. The loose subcutaneous tissue contains numerous elastic fibres but no fat cells as a rule. Two or three

rows of stiff hairs, the cilia or eyelashes, occur at the margin, and are shed and renewed about once in four months. Into the follicles of the cilia, open sebaceous glands and the glands of Moll, which are modified sweat glands.

2. The muscle layer of striated muscles of the orbicularis palpebrarum, lying transversely, adjoins the subcutaneous tissue; the ciliary or marginal muscle is a portion of this muscle at the posterior margin of the eyelid.

3. The connective-tissue layer consist of areolar tissue, the fascia palpebralis, to which some of the fibres of the tendon of the levator palpebrae pass; others are attached to the upper portion of the next layer, the tarsus, and contain smooth muscle-fibres, the lid-muscle of Müller.

4. The tarsus, a plate of dense fibrous tissue, forms a supporting structure to the eyelid and partially encloses the Meibomian glands, which are modified sebaceous glands, extending through two-thirds of the height of the eyelid from the posterior free margin, where the long excretory duct opens. In structure they resemble sebaceous glands. Above the tarsus in the nasal half of the eyelid are branched tubular glands, the accessory lachrymal glands.

5. The conjunctiva, which lines the inner surface of the eyelid, consists of a tunica propria, composed of fibrous tissue, plasma and lymphoid cells, and covered by stratified columnar epithelium. The part of the conjunctiva beyond the tarsus is folded, and in sections the depressions resemble glands. The palpebral conjunctiva and the scleral conjunctiva meet along the fornix conjunctivae, where a number of small lymph-nodules and mucous glands occur. The epithelium of the scleral conjunctiva becomes stratified squamous as it approaches the cornea.

The plica semilunaris, a rudimentary third eyelid, the membrana nictitans of some animals, is a fold of the conjunctiva, having a stratified squamous epithelium.

The caruncula lachrymalis, an isolated portion of the integument, consists of epidermis without the stratum corneum, and a corium and subcutaneous tissue, with fat-cells, modified sweat glands, hairs, and smooth and sometimes striated muscle-fibres. The blood-vessels, coming from the outer and the inner angles, form the arcus tarseus at the margin of the lid, and the arcus tarseus externus at the upper margin of the tarsus. From these arches, branches pass to the various structures of the eyelid and form a capillary plexus beneath the palpebral conjunctiva, supplying also the fornix conjunctivae and the scleral conjunctiva.

The lymphatics form a network in the conjunctiva and another on the anterior surface of the tarsus. The two systems are in communication by vessels passing through the tarsus.

The nerves of the eyelid form a rich plexus at the margin.

IV. The lachrymal apparatus consists of the lachrymal gland, the lachrymal canaliculi, the lachrymal sac, and the naso-lachrymal duct.

1. The lachrymal gland is a compound tubular serous gland. The numerous excretory ducts, lined with two layers of columnar epithelium, pass into intercalated tubules, lined with cuboidal cells, and these pass to the terminal tubules, which have a simple columnar epithelium.

2. The lachrymal canaliculi are lined with stratified squamous epithelium, resting on a tunica propria of fibrous tissue and numerous elastic fibres circularly arranged; outside is a longitudinal layer of striated muscle-fibres.

3. The lachrymal sac and naso-lachrymal duct are lined with two layers of columnar epithelium resting on a tunica propria chiefly of adenoid tissue, which is separated from the adjacent periosteum by areolar tissue containing a rich venous plexus.

B. THE AUDITORY APPARATUS.

The ear consists of three parts, the external ear, the middle ear, and the internal ear.

I. The external ear.

1. The pinna and a part of the external auditory canal consist of a framework of yellow, elastic cartilage, covered with integument and subcutaneous tissue, in which hairs and sebaceous glands are present and also the ceruminous glands, which are modified sweat glands. These glands consist of coiled tubules lined with a single layer of cuboidal cells, which rest on a basement membrane enclosed in a thin longitudinal layer of smooth muscle-fibres. The excretory duct is lined with a stratified epithelium of cuboidal cells.

2. The external auditory canal is for the most part a bony canal lined with a thin integument adhering closely to the periosteum and devoid of hairs and glands.

3. The tympanic membrane consists of three layers:

a. The integument covers the outer surface and consists of epidermis and a thin corium.

b. The lamina propria consists of fibrous tissue, whose fibres are arranged in two layers, an outer radial, and an inner circular.

c. The mucous layer covers the inner surface of the lamina propria and consists of a thin tunica propria covered by a simple cuboidal non-ciliated epithelium. Where the malleus is attached to the tunica propria a layer of hyaline cartilage is present.

The blood-vessels have the usual distribution in the integument; in the mucous layer is a capillary network connected with the vessels of the malleus; branches pass from the mucous layer through the lamina propria and anastomose with vessels in the integument. Lymphatics and nerves occur in the integument and in the mucous layer.

II. The middle ear.

1. The tympanic cavity is lined with a mucous membrane, consisting of a thin tunica propria closely attached to the periosteum, and an epithelium. On the tympanic membranes, the promontory, the ear-ossicles, and in the antrum and the mastoid cells, the epithelium is simple cuboidal non-ciliated; in the rest of the tympanic cavity it is ciliated columnar, with a layer of small cells at the base of the columnar cells.

2. The secondary tympanic membrane fills the fenestra rotunda and consists of three layers :

a. The mucous layer, the outermost;

b. The fibrous layer, consisting of radially arranged bundles of fibrous tissue;

c. The innermost layer, of fibrous tissue covered with endothelium.

3. The ear-ossicles consist of typical compact bone; the ends of the bones are covered with hyaline cartilage, and the mucous membrane of the tympanic cavity covers the free surfaces.

4. The Eustachian tube is supported in part by bone, in part by cartilage, and in part by fibrous tissue. It is lined with a mucous membrane having in some parts a loose, fibro-elastic submucosa. The mucous membrane consists of stratified ciliated columnar epithelium, resting on a fibrous tunica propria containing lymphoid cells. Small mucous glands are present and are especially numerous near the pharyngeal portion of the tube.

The blood-vessels form a superficial capillary network in the mucosa and a deeper network around the glands. The nerves form a similarly situated plexus; their termination is not well known.

III. The internal ear consists of a cavity in the sphenoid, the bony labyrinth, lined with a thin, fibrous periosteum, whose surface is covered with endothelium; within the bony labyrinth lies the membranous labyrinth, consisting of the sacculus, the utriculus, the semicircular canals and the membranous cochlea. The membranous labyrinth is separated from the bony labyrinth by a space filled with a fluid, the **perilymph**; within the membranous labyrinth is a similar fluid, the **endolymph**. The utriculus and the sacculus are connected by the ductus endolymphaticus. The utriculus is connected with the semicircular canals, each of which has an enlargement, the **ampulla**, at the point of opening. The sacculus connects with the membranous cochlea through the canalis reuniens.

1. The **utriculus**, the **sacculus** and the **semicircular canals** are essentially alike in structure. The walls consist of three layers : —

a. The outermost or **fibrous layer** consists of fibrous and elastic tissue and is connected by trabeculae with the periosteum of the bony labyrinth ; it contains a network of capillary blood-vessels.

b. The **basement membrane** separates the outer from the inner layer and exhibits elevations in places on its inner surface.

c. The **epithelial layer** consists of simple squamous epithelium except at the points where groups of nerve-fibres terminate, *i. e.* on the **maculae acusticae** of the utricle and the saccule, and the **cristae acusticae** in the ampullae of the semicircular canals. In these areas the fibrous layer and the basement membrane are thickened, while the epithelium becomes first columnar and then, on the maculae and the cristae themselves, **neuro-epithelium**. This consists of **fibre-cells**, which are sustentacular cells extending the entire thickness of the epithelium and expanded at each end, and **hair-cells**, which are cylindrical and extend from the free surface of the epithelium through half its thickness ; at the surface they possess several, long agglutinated processes, the **auditory hairs**. The **nerve-fibres** enter the epithelium as naked axis-cylinder processes and divide at the base of the hair-cells into several, varicose branches, which run parallel to the surface and send off occasional twigs, which terminate in free ends on the lateral surfaces of the hair-cells. A **cuticula** covers the free surface of the neuro-epithelium and the adjacent columnar epithelium. The **otolith membrane** lies on the free surface of the maculae ; it is gelatinous and contains many hexagonal prisms of calcium carbonate, the **otoliths**, from 1 to 15 μ in length. The **cupola** lies on the cristae and is not apparent until coagulated by some fixing reagent.

2. The **cochlea** may be oriented conveniently with the apex uppermost, considering the axis or modiolus, the inner portion of the organ, — the periphery, the outer portion. From the axis a bony plate, the **bony spiral lamina**, extends horizontally into

the middle of the tubular lumen of the cochlea. The **membranous cochlea** or **ductus cochlearis**, is triangular in cross section, with the base of the triangle attached to the peripheral wall, and its apex resting on the peripheral free margin of the bony spiral lamina. The large **perilymphatic space** above the ductus cochlearis is the **scala vestibuli**, the similar space below is the **scala tympani.** The membrane between the ductus cochlearis and the scala vestibuli is the **membrane of Reissner**, or **membrana vestibularis**, that between the ductus cochlearis and the scala tympani is the **membranous spiral lamina.** The periosteum which lines the scala vestibuli and the scala tympani is greatly thickened on the peripheral side of the cochlea, forming a crescentic mass, the **ligamentum spirale**, which is attached to the ductus cochlearis and extends both above and below it.

a. The **ligamentum spirale**, consisting of fibrous tissue, possesses a projection on its free surface, the **crista basilaris**, to which the membranous spiral lamina is attached. Above this there is a smaller projection, into the cavity of the ductus cochlearis ; this is the **prominentia spiralis** containing a bloodvessel, the **vas prominens.** Between the prominentia spiralis and the membrane of Reissner is a band of tissue containing a close network of capillary blood-vessels, the **stria vascularis**, lying just beneath the epithelium. On the stria vascularis is a **simple cuboidal epithelium**, which becomes flattened on the prominentia, and then cuboidal again near the membranous spiral lamina.

b. The **membrane of Reissner** is very thin and consists of three layers : — an **endothelium** adjoining the scala vestibuli, a middle layer of **fibrous connective tissue**, and a **simple squamous epithelium** lining this portion of the ductus cochlearis.

c. The **tympanic wall** consists of (1), the **limbus**, which extends from the lower attachment of the membrane of Reissner to the free edge of the bony spiral lamina, and (2), the **membranous spiral lamina.**

The **limbus** consists of a thick mass of connective tissue continuous with the periosteum, and terminates peripherally in an upper and a lower projection with a recess or groove between them, constituting the **superior labium** or **labium vestibulare**, the **inferior labium** or **labium tympanicum**, and the **sulcus spiralis.** The upper surface of the limbus exhibits numerous irregular elevations, which on the labium vestibulare take the form of elongated plates arranged at right angles to the free margin and constituting the **auditory teeth**; this surface is covered with simple flattened epithelium continuous with that of the membrane of Reissner, but in the grooves between the teeth it becomes columnar. The sulcus spiralis is covered with low cuboidal epithelium. Near the peripheral margin of the upper surface of the bony spiral lamina is a region, the zona perforata, marked by a row of cleft-like openings, the foramina nervina, through which nerve-fibres pass to the epithelium of the membranous spiral lamina.

The membranous spiral lamina consists of three layers : — (1) the tympanic lamella is adjacent to the scala tympani, and consists of fibres and spindle-shaped cells lying at right angles to the fibres of the layer above ; there is no true endothelium ; (2) the basilar membrane or membrana propria is a firm membrane exhibiting fine parallel fibres extending from the bony spiral lamina to the crista basilaris of the ligamentum spirale ; (3) the epithelium resting on the basilar membrane forms the floor of the ductus cochlearis ; the inner portion consists of neuro-epithelium constituting the spiral organ or organ of Corti, — the outer portion consists of low columnar cells, the cells of Claudius. That part of the spiral lamina covered by the organ of Corti is known as the zona tecta, that part covered by the cells of Claudius as the zona pectinata.

The organ of Corti extends through the entire length of the cochlea excepting a short distance at each end. It possesses an inner and an outer row of firm, supporting cells, the pillar cells or rods of Corti, which rest on the basilar membrane, while their upper ends meet; thus they form the arcus spiralis and enclose a triangular canal, the tunnel of Corti. This tunnel is an intercellular space, containing a semi-fluid substance, and is traversed by nerve-fibres.

The pillar-cells possess expanded bases, triangular in section, narrow bodies, and heads expanded into thin head plates directed outwards. The heads of the inner pillar-cells are concave on their outer surfaces ; the heads of the outer pillar-cells are convex on their inner surfaces, and their head plates extend a short distance beyond the free edges of the overlying plates of the inner cells. A distinct oval area occurs in each head near the curved surface of contact. A small amount of protoplasm encloses the body and the base of each pillar-cell and contains a nucleus at the base, on the side turned towards the tunnel. The inner pillars are more numerous than the outer.

The inner hair-cells constitute a single row of short cells on the inner side of the inner pillar-cells. The hair-cells are less numerous than the pillar cells. They are from one-third to one-half the height of the inner pillar-cells, and each bears on its free surface some twenty fine processes or hairs. On the inner side of these cells are supporting epithelial cells, which are continuous with the epithelium of the sulcus spiralis.

The outer hair-cells constitute three or four rows, the cells in adjacent rows alternating with one another. In structure these cells are similar to the inner hair-cells; a dark structure, the spiral body, lies in the upper portion of each cell. They are separated from one another by firm supporting cells, the cells of Deiters, which rest on the basilar membrane and possess cuticular, expanded, upper ends, the phalanges, which are united with one another and with the head plates of the outer pillar-cells, forming the membrana reticularis. Below the outer hair-cells, intercellular spaces exist between the cells of Deiters; they are known as Nuel's spaces and communicate with one another, and also with the tunnel by means of the clefts between the bodies of the outer pillar-cells.

The cells of Hensen are elongated epithelial cells adjoining the outermost row of Deiters' cells and become continuous peripherally with the cells of Claudius.

The membrana tectoria is a cuticular structure containing fine fibres, which is attached to the edge of the labium vestibulare and extends over the organ of Corti as far as the outermost row of outer hair-cells.[1]

The nerves arise from the cochlear branch of the auditory nerve; they lie in the axis of the cochlea and send off branches continually, which pass into the ganglion spirale, a continuous winding ganglion lying in the base of the bony spiral lamina. The ganglion-cells are bipolar. From the ganglion the nerve-fibres go to the zona perforata, where they lose their medullary sheath and pass through the foramina nervina. Bundles of fibres then pass to the inner hair-cells; others pass between the inner pillar-cells and across the tunnel to the several rows of outer hair-cells; fine fibres terminate in connection with the hair-cells, — the method of termination is still in dispute.

The arteries arise from the cochlear branch of the auditory artery; this branch divides, and some of its branches go directly to the initial portion of the cochlea, while others enter the modiolus where they form masses of coiled vessels, the small and the large glomeruli cochleae. From one set of glomeruli branches pass to Reissner's membrane and to the limbus; from the other, to the membranous spiral lamina and to the stria vascularis. The capil-

---

[1] According to H. Ayers (Jour. of Morph., 1892), the membrana tectoria consists merely of the elongated hairs of the hair-cells, and is consequently attached to the organ of Corti and lies above or upon the labium vestibulare but not fused with it.

laries connect with the vas prominens and with the vas spirale, which lies beneath the basilar membrane in the region of the tunnel of Corti, and these veins open into the vena spiralis modioli, which lies beneath the ganglion spirale.

The **perilymphatic spaces** connect with the subarachnoid spaces ; the **endolymph** is in communication with the subdural spaces.

C. THE NASAL MUCOUS MEMBRANE.

I. The **vestibular region** is lined with a modified integument, consisting of stratified squamous epithelium, resting on a tunica propria having papillae and enclosing sebaceous glands and hair follicles.

II. The **respiratory region** exhibits a thick mucosa, often 4 mm. It consists of stratified ciliated columnar epithelium, which contains numerous goblet-cells, and a tunica propria of fibrous tissue containing numerous leucocytes and some simple lymph nodules. Mixed tubular glands are numerous, — having a mucous and serous secretion. Veins are so numerous in some regions as to suggest the structure of cavernous tissue.

III. The **olfactory region** is yellowish-brown in colour, instead of red like the respiratory region. It possesses an epithelium and a tunica propria.

The epithelium exhibits three kinds of cells : —

1. The **supporting cells** have a broad outer end, an oval nucleus, and usually an attenuated, irregular or forked, inner end. They contain granules of yellowish-brown pigment arranged more or less regularly in vertical rows.

2. The **olfactory cells** are nerve-cells, having a slender outer process terminating in a ciliated end ; an enlargement contains the spherical nucleus, and the inner end is continuous with a non-medullated nerve-fibre. There are several layers of these cells.

3. The **basal cells** form a layer of small cells adjoining the tunica propria.

The free surface of the olfactory epithelium possesses a thin membrane, the **membrana limitans olfactoria**; the supporting cells are generally considered non-ciliated.

The **tunica propria** consists of fibrous and elastic tissue and contains numerous branched, tubular, mucous glands, the **glands of Bowman**. The **blood-vessels** form a rich subepithelial capillary network ; the veins are very numerous. **Lymph-vessels** and **lymph-clefts** are numerous in the deeper parts of the tunica propria. In addition to the **olfactory nerves** connected with the olfactory cells, medullated branches from the trifacial nerve occur in the tunica propria.

INDEX.

www.ingramcontent.com/pod-product-compliance
Lightning Source LLC
Chambersburg PA
CBHW021946190326
41519CB00009B/1157